T0182101

Mathematical Modeling of Mitochondrial Swelling

Messoud Efendiev

Mathematical Modeling
of Mitochondrial Swelling

 Springer

Messoud Efendiev
Institute of Computational Biology
Helmholtz Center Munich
Neuherberg, Germany

ISBN 978-3-030-07567-5   ISBN 978-3-319-99100-9   (eBook)
https://doi.org/10.1007/978-3-319-99100-9

Mathematics Subject Classification (2010): 35B40, 35K45, 37N25

This Springer imprint is published by the registered company Springer Nature Switzerland AG
The registered company address is: Gewerbestrasse 11, 6330 Cham, Switzerland

*"Mieux vaut une tête bien faite qu'une tête
bien pleine"*
  *"Better a well-made mind than
a well-filled mind"*
  (Michel Montaigne)
  *Of course, very few people today enjoy
both, and they are therefore exceptional,
and very important for mankind. One
of the people whom I have met in my life who
has both completely is Sadik Murtuzayev.
This book is dedicated to him with admiration
on the occasion of his 90th birthday.*

Preface

In this book, new mathematical models of mitochondria swelling are presented which take into account, in particular, spatial effects. Depending on the models under considerations, the problem consists of either one reaction diffusion (including degenerate diffusion) equation coupled with a system of three ordinary differential equations (ODEs), describing the evolution in time of three subpopulations of mitochondria, or of four coupled partial differential equations (PDEs) between calcium concentration and subpopulations of mitochondria.

The models in the book are of the form

$$\partial_t u = d_1 A(u) + d_2 g(u) N_2,$$
$$\partial_t N_1 = d_3 \Delta_x N_1 - f(u) N_1,$$
$$\partial_t N_2 = d_4 \Delta_x N_2 + f(u) N_1 - g(u) N_2,$$
$$\partial_t N_3 = d_5 \Delta_x N_3 + g(u) N_2.$$

where the coefficients d_1, d_2 are positive constants, while d_3, d_4, and d_5 are nonnegative constants and f and g are given functions, describing the transition from unswollen $N_1(x, t)$ to swollen $N_2(x, t)$ and from swollen $N_2(x, t)$ to completely swollen $N_3(x, t)$ mitochondria.

The boundary conditions vary for the type of the model and can be generally formulated as $a(x)u + b(x)\partial_\nu u = h(x)$ on the boundary of Ω, reflecting the swelling of mitochondria scenarios in vitro and in vivo, where a, b, h are given real functions that are defined on the boundary of Ω and $\partial_\nu u$ is the normal derivative at the boundary. As mentioned above the novelty in this book compared with existing mitochondria models is that it takes into account spatial effects by means of the diffusion operator A both with and without spatial diffusion for mitochondria subpopulations. In the book, we consider both the nondegenerate diffusion $A(u) = \Delta_x u$, where Δ_x is the standard Laplacian, and the degenerate case $A(u) = \Delta_x u^m$ with $m > 1$.

Here Ω is a bounded domain in $\mathbb{R}^n$, which will be, depending on the model, either the test tube (in vitro) or the whole cell (in vivo).

We especially emphasize that the new approach developed in this book is capable of modeling each of the in vitro, in silico, and in vivo cases. The cases vary with respect to the initial and boundary conditions dependent on the choice of the domain as either the test tube or the whole cell.

The book consists of seven chapters. Chapter 1 is of an introductory character and includes four subsections. In Sect. 1.1, we give very briefly the definitions of Sobolev spaces and we state the embedding theorems. Section 1.2 is devoted to the generalization of Poincare inequalities. In Sects. 1.3 and 1.4, we give the properties of Nemytski operators and the subdifferential in Hilbert spaces respectively which play a crucial role in our analysis of the mitochondria models.

Chapter 2 deals with the biological background of mitochondria swelling and its role in the apoptosis scenario.

In Chap. 3, we study a model of the description of mitochondria swelling based on experimental observations. Chapter 3 consists of four subsections: Sect. 3.1 is devoted to existing models that are focused on the evolution in time of mitochondria and, as a consequence, are described by ordinary differential equations (ODE models); they do not take into account local effects and only work with mean values over the whole domain. Coming from an ODE model, it is natural to think about the necessity of including spatial effects by means of considering partial differential equations (PDE models). We consider spatial effects in Sect. 3.2. Moreover, we show in Sect. 3.2 that how to take into account this spatial effect is dictated by experiments. Indeed, in this section it is shown that the same amount of mitochondria with different concentrations has an influence on the volume outcome, that is, leads to different mitochondria swelling scenarios. With the necessity of taking into account spatial effects, in Sect. 3.3 we deal with the mitochondria model in vitro. We assume that mitochondria in the test tube as well as within the cell do not move in any direction, and hence, the spatial effects are only introduced by the calcium evolution. As a consequence of this assumption, our mathematical model of mitochondria swelling is described by PDE-ODE systems (the so-called PDE-ODE coupling) with Neumann boundary conditions. We emphasize that in this vitro model we choose the underlying domain to be the test tube.

Chapter 4 is devoted to the mathematical analysis of the model obtained in Sect. 3.3 and consists of four subsections. In Sect. 4.1, we verify the obtained model and prove the well posedness of the PDE-ODE systems corresponding to it. Having ordinary differential equations describing the evolution in time of three subpopulations of mitochondria indicates that we assume the mitochondria in the test tube, as well as within cells, do not move in any direction and hence the spatial effects are only introduced by the calcium evolution obeying a partial differential equation (PDE), namely a reaction-diffusion equation. In Sect. 4.2, we study the long-time dynamics of solutions of the PDE-ODE coupling. We obtain the complete classifications of the limiting profile of solutions and study partial and complete classification scenarios depending on the given data. Note that these scenarios, namely, partial and complete swelling scenarios, have been observed in experiments. Section 4.3 deals with the numerical simulation (in silico) of PDE-ODE systems. One of the remarkable results here is the clearly visible spreading calcium wave. If we compare the dynamics with those of simple diffusion without any positive feedback, the numerical results show that the resulting calcium evolution induced by mitochondria swelling is indeed completely different. Our numerical simulations show that a small change in the initial distribution of calcium is enough to shift the behavior from partial to complete swelling behavior. In Sect. 4.4, we continue our analysis of a coupled PDE-ODE model of calcium-induced mitochondria swelling in vitro. More precisely, we study the long-time dynamics of solutions of PDE-ODE systems under homogeneous Dirichlet boundary conditions. Note that, biologically, this kind of boundary condition appears if we put some chemical material on the test-tube wall that binds calcium ions and hence removes it as a swelling inducer. We especially emphasize that the analytical machinery that was developed in Sect. 4.1 is not applicable under homogeneous Dirichlet boundary conditions and therefore must be extended. In this section, we show that the calcium ion concentration will tend to zero and that, in general, complete swelling will not occur as time goes to infinity. This distinguishes the situation under Dirichlet boundary conditions from the situation under Neumann boundary conditions that were analyzed in Sect. 4.1. In Sect. 4.5, we carry out numerical simulations validating the analytical results of Sect. 4.4.

Chapter 5 deals with the swelling process in a living organism, so that in this case we do not have a controlled environment as we had in the test tube. Here there are three main factors that differ a great deal from the in vitro case, which was considered in Chap. 3, that is:

1. Mitochondria are not uniformly distributed.
2. The induced calcium source is very localized.
3. The cell is not a closed system.

Chapter 5 consists of two subsections. In Sect. 5.4, we formulate mitochondria swelling scenario in vivo as a PDE-ODE system with Robin boundary conditions, which in turn reflects the fact that calcium ions can enter and leave the cell across the permeable membrane. In this section, we prove the well posedness and study large-time asymptotics of solutions of the corresponding initial boundary value PDE-ODE problem. Moreover, in Sect. 5.4 it is shown that partial and complete swelling behavior in vivo depends on the balance of concentration as well as on the threshold parameter of initiation for swelling. In Sect. 5.5, we illustrate the results of Sect. 5.4 with numerical simulations. Note that in the previous chapters mathematical modeling of swelling of mitochondria and their analysis in vitro, in vivo, and in silico were governed by the concept of standard diffusion for calcium concentrations. The use of the standard Laplacian to describe diffusion processes is generally accepted for mathematical idealization and indeed can explain some aspects of mitochondria swelling scenarios observed in experiments. However, from the viewpoint of applications it is also desirable to analyze the effect of degenerate diffusion, because there is no species which can diffuse infinitely fast (a property of the standard diffusion). Therefore in Chap. 6, which consists of two subsections, we consider the effect of degenerate diffusion on mitochondria swelling behavior. In Sect. 6.1, we study the well posedness of solutions under homogeneous Dirichlet boundary conditions. Note that we cannot apply the methods developed in previous chapters for nondegenerate diffusion and new methods have been developed in order to study both well posedness and asymptotic behavior of solutions. In this section, we show that the calcium ion concentration will tend to zero and that, in general, complete swelling will not take place as time goes to infinity in the case of degenerate diffusion. Section 6.2 gives numerical simulations.

Note that in Chaps. 1–6 we essentially used the assumption that mitochondria do not diffuse within the cell, which in turn led to ordinary differential equations for the evolution of mitochondria subpopulations N_i, $i = 1, 2, 3$, as a result of PDE-ODE coupling between calcium concentration and subpopulations of mitochondria. However, there are indications that mitochondria do move under certain circumstances in the test tube as well as within the cell in any direction, e.g., dependent on the cell cycle, which in turn leads to partial differential equations for mitochondria subpopulations $N_i(x, t)$.

Chapter 7 consists of three sections. In Sect. 7.1, we systematically study this PDE-PDE case as for the well posedness, dependence of solutions from boundary conditions (in vitro and in vivo cases). In Sect. 7.2, we study asymptotic behavior of solutions as for the partial and complete swelling scenarios. Section 7.3 verifies the obtained results numerically.

Note that the development and study of mathematical models that take into account spatial effects for the swelling of mitochondria was initiated by myself in the interdisciplinary seminar at the Helmholtz Center of Munich and was deepened afterward by the numerous, intensive, and stimulating discussions with my colleagues Dr. H. Zischka and Dr. S. Schulz from the Institute of Toxicology and Pharmacology at the Helmholtz Center of Munich.

I emphasize that the mathematical analysis of the obtained model was first carried out in collaboration with my Ph.D. student S. Eisenhofer, where, in particular, several challenging and open questions were formulated that must be answered in order to shed light on and gain a deeper understanding of mitochondria swelling scenarios in vitro, in vivo, and in silico. Together with my frequent coauthors and friends, that is, with Prof. M. Otani (Japan) and Prof. H. Eberl (Canada), during the last few years we systematically studied many of these challenging and open questions. Therefore, I express my sincere thanks to all of my colleagues and friends mentioned above, because without such stimulating and inspiring discussions and joint works this book would not be possible.

I also thank the many friends and colleagues who gave me suggestions, advice, and support. In particular, I wish to thank professors J.R.L. Webb, J. Wu, F. Hamel, Y. Nishiura, M. Sörensen, and M. Pedersen.

Furthermore, I am greatly indebted to my colleagues at the Helmholtz Zentrum München and Technische Universität München, the Alexander von Humboldt Foundation, as well as Springer Publisher for their efficient handling of the publications.

Last but not least, I wish to thank my family for continuously encouraging me during the writing of this book.

Neuherberg, Germany   Messoud Efendiev

Contents

Chapter 1
Functional Spaces

1.1 Sobolev Spaces and Embeddings Theorem

We shall use the following notation. We shall denote by $\mathbb{R}$, $\mathbb{C}$, $\mathbb{Z}$, and $\mathbb{N}$ the sets of real, complex, integer, and natural numbers, respectively; $\mathbb{Z}_+ = \{x \in \mathbb{Z} \mid x \geqslant 0\}$ is the set of nonnegative integers. $\mathbb{R}^n$ is the standard real vector space of dimension n. We denote by D_i the operator of partial differentiation with respect to x_i:

$$D_i u = \frac{\partial u}{\partial x_i} \quad (i = 1, \ldots n).$$

As usual, we use multi-index notation to denote higher order partial derivatives:

$$D^\gamma = D_1^{\gamma_1} \cdots D_n^{\gamma_n}, \quad |\gamma| = \gamma_1 + \cdots + \gamma_n$$

is a partial derivative of order $|\gamma|$ for a given $\gamma = (\gamma_1, \ldots, \gamma_n)$, $\gamma_i \in \mathbb{Z}_+$.

Let $u : \Omega \subset \mathbb{R}^n$ be a real function defined on a bounded domain Ω. The space of continuous functions over the bounded domain $\bar{\Omega}$ is denoted by $C(\bar{\Omega})$; the norm in $C(\bar{\Omega})$ is defined in a standard way:

$$\|u\|_{C(\bar{\Omega})} = \sup\{|u(x)| \mid x \in \bar{\Omega}\}.$$

The space $C^m(\Omega)$ consists of all real functions on Ω which have continuous partial derivatives up to order m. By definition, u belongs to $C^m(\bar{\Omega})$ iff (abbreviation for if and only if) $u \in C^m(\Omega)$ and u and all its partial derivatives up to order m can be extended continuously to $\bar{\Omega}$. Let $0 < \gamma < 1$ and $k \in \mathbb{Z}_+$. By definition $C^{k,\gamma}(\bar{\Omega})$

© Springer Nature Switzerland AG 2018
M. Efendiev, *Mathematical Modeling of Mitochondrial Swelling*,
https://doi.org/10.1007/978-3-319-99100-9_1

denotes the Hölder space of functions $u : \Omega \to \mathbb{R}$ such that $D^\alpha u : \Omega \to \mathbb{R}$ exists and is uniformly continuous when $|\alpha| = k$ and

$$|u|_{k,\gamma} \equiv \sup \left\{ \frac{|D^\alpha u(x) - D^\alpha u(y)|}{|x - y|^\gamma} \mid x, y \in \Omega,\ x \ne y,\ |\alpha| \leqslant k \right\} \tag{1.1}$$

is finite. For $u \in C^{k,\gamma}(\bar{\Omega})$, we set

$$\|u\|_{k,\gamma} = |u|_{k,\gamma} + \sum_{|\alpha| \leqslant k} \max\{|D^\alpha u(x)| \mid x \in \bar{\Omega}\}.$$

We also introduce

$$C^{k,\gamma}(\partial\Omega) = \{\varphi : \partial\Omega \to \mathbb{R} \mid \text{there exists } u \in C^{k,\gamma}(\bar{\Omega}) \text{ with } u|_{\partial\Omega} = \varphi\},$$

and for $\varphi \in C^{k,\gamma}(\partial\Omega)$ we set

$$\|\varphi\|_{k,\gamma} = \inf \left\{ \|u\|_{k,\gamma} \mid u|_{\partial\Omega} = \varphi;\ u \in C^{k,\gamma}(\bar{\Omega}) \right\}.$$

In cases when it is clear from the context where the function under consideration is defined, we shall sometimes simply write $u \in C^k$ instead of, for example, $u \in C^k(\mathbb{R}^n)$. In several examples we shall use the spaces of functions that are 2π-periodic in every variable x_i ($i = 1, \ldots, n$). We shall consider such functions as being defined on the n-dimensional torus $T^n = \mathbb{R}^n/(2\pi\mathbb{Z})^n$. We denote by $L^p(\Omega)$, $1 \leqslant p \leqslant \infty$, the space of measurable functions with the finite norm

$$\|u\|_{0,p} = \|u\|_{L^p(\Omega)} = \left(\int_\Omega |u(x)|^p dx \right)^{1/p}. \tag{1.2}$$

We denote by $L^\infty(\Omega)$ the space of almost everywhere bounded functions,

$$\|u\|_{0,\infty} = \|u\|_{L^\infty} = \text{vrai}\sup\{|u(x)| \mid x \in \Omega\}$$

(for continuous functions this norm coincides with the norm of $C(\bar{\Omega})$). The norm in the Sobolev space $W^{l,p}(\Omega)$, $l \in \mathbb{Z}_+$, $1 \leqslant p < \infty$, is defined by the formula

$$\|u\|_{l,p} = \left(\sum_{|\alpha| \leqslant l} \|D^\alpha u\|_{L^p(\Omega)}^p \right)^{1/p}. \tag{1.3}$$

In the case $p = 2$ this Sobolev space is a Hilbert space and is denoted $H^l(\Omega)$, i.e., $H^l(\Omega) = W^{l,2}(\Omega)$. The scalar product in $H^l(\Omega)$ is defined by the formula

$$(u, v)_l = \sum_{|\alpha| \leqslant l} \int_\Omega D^\alpha u(x) \cdot D^\alpha v(x) dx. \tag{1.4}$$

The space $W^{l,p}(\Omega)$ is the completion of $C^l(\Omega)$ with respect to the norm (1.3). The norms $C^{k,\gamma}(T^n)$ and $W^{l,p}(T^n)$ are defined by (1.1) and (1.3) with $\Omega = (0, 2\pi)$. The scalar product and the norm in $H^l(T^n)$, which are equivalent to those defined by (1.4), are defined in terms of Fourier coefficients,

$$(u, v)_l = \sum \tilde{u}(\xi) \cdot \overline{\tilde{v}(\xi)} \cdot (1 + |\xi|^2)^l; \quad \|u\|_l^2 = \langle u, u \rangle_l, \tag{1.5}$$

where the summation is over $\xi \in \mathbb{Z}^n$; the bar denotes complex conjugation; $\tilde{u}(\xi)$ and $\tilde{v}(\xi)$ are the Fourier coefficients:

$$\tilde{u}(\xi) = (2\pi)^{-n} \int_\Omega u(x) e^{-ix \cdot \xi} dx.$$

Here the integration is over $[0, 2\pi]^n$ and

$$x \cdot \xi = x_1 \xi_1 + \cdots + x_n \xi_n.$$

The formula (1.5) defines a norm in $H^l(T^n)$ not only for $l \in \mathbb{Z}_+$ but for $l \in \mathbb{R}$ as well. We denote by $C^\infty(\bar{\Omega})$ the space $\bigcap_{k \geqslant 0} C^k(\bar{\Omega})$; by $C_0^\infty(\Omega)$ the set of functions from $C^\infty(\bar{\Omega})$ which vanish on a neighborhood of the boundary $\partial\Omega$. We shall use also spaces of functions which vanish on $\partial\Omega$. In this case we shall denote the corresponding space as follows:

$$C^{k,\gamma}(\bar{\Omega}) \cap \{u|_{\partial\Omega} = 0\}, \quad W^{l,p}(\Omega) \cap \{u|_{\partial\Omega} = 0\}.$$

We denote the completion of $C_0^\infty(\Omega)$ with respect to the norm of $H^l(\Omega)$ by $H_0^l(\Omega)$ and with respect to the norm of $W^{1,p}(\Omega)$ by $W_0^{1,p}(\Omega)$. It is well known that

$$H_0^1(\Omega) = H^1(\Omega) \cap \{u|_{\partial\Omega} = 0\}; \quad W_0^{1,p}(\Omega) = W^{1,p} \cap \{u|_{\partial\Omega} = 0\}.$$

The Sobolev spaces $H^\rho(\Omega)$ with non-integer $\rho \geqslant 0$, $\rho = k + \beta$, $k \in \mathbb{Z}$, $0 \leqslant \beta < 1$ are endowed with the norm

$$\|u\|_\rho^2 = \|u\|_k^2 + \int_{|y| \leqslant \delta} \|u(x + y) - u(x)\|_k^2 \cdot |y|^{-n-2\beta} dy$$

(u is extended over a δ-neighborhood of the boundary, see [6]).

By $S(\mathbb{R}^n)$ we denote the class of rapidly decreasing (at ∞) functions $u \in C^{\infty}(\mathbb{R}^n)$, with

$$(1 + |x|)^k \, |D^\alpha u(x)| \leqslant C_{k,\alpha}$$

for each $\alpha = (\alpha_1, \cdots , \alpha_n) \in \mathbb{Z}_+^n$ and $k \in \mathbb{Z}_+$, where $C_{k,\alpha}$ are constants.

Recall that an operator $j : X \to Y$ between Banach spaces with $X \subseteq Y$ is an embedding iff $j(x) = x$ for all $x \in X$. The operator j is continuous iff $\|x\|_Y \leqslant$ constant $\|x\|_X$ for all $x \in X$. Further, j is compact iff j is continuous, and every bounded set in X is relatively compact in Y. If the embedding $X \hookrightarrow Y$ is compact, then each bounded sequence $\{x_n\}$ in X has a subsequence $\{x_{n'}\}$ which is convergent in Y.

We will widely use the Sobolev embedding theorems formulated below.

Theorem 1.1 *Let Ω be a bounded domain in $\mathbb{R}^n$, with smooth boundary $\partial\Omega$ and $0 \leqslant k \leqslant m - 1$. (See [1, 6, 11].) Then*

$$W^{m,p}(\Omega) \hookrightarrow W^{k,q}(\Omega), \text{ if } \frac{1}{q} \geqslant \frac{1}{p} - \frac{m-k}{n} > 0$$

$$W^{m,p}(\Omega) \hookrightarrow W^{k,q}(\Omega), \text{ if } q < \infty, \ \frac{1}{p} = \frac{m-k}{n}$$

$$W^{m,p}(\Omega) \hookrightarrow C^{k,\delta}(\bar{\Omega}), \text{ if } \frac{n}{p} < m - (k+\delta), \ 0 < \delta < 1.$$

The last two embeddings are compact, the first embedding is compact if $\frac{1}{q} > \frac{1}{p} - \frac{m-k}{n}$.

Theorem 1.2 *Let $0 \leqslant \beta < \alpha \leqslant 1$ or $\alpha, \beta \in \mathbb{Z}$ with $0 \leqslant \beta < \alpha$ (see [7, 11]). Then the embedding*

1)

$$C^\alpha(\bar{\Omega}) \hookrightarrow C^\beta(\bar{\Omega}) \text{ is compact}$$

and

2) for $k + \beta < m + \alpha$, with $0 \leqslant \alpha, \beta \leqslant 1$, $m \geqslant k \geqslant 0$ the embeddings

$$C^{m,\alpha}(\bar{\Omega}) \hookrightarrow C^{k,\beta}(\bar{\Omega})$$

are compact.

The following Lemma 1.1 is of independent interest. Since this is not known to a broader audience we present below its proof.

Let Ω be a bounded domain in $\mathbb{R}^3$ and let $a \in L^\infty(\Omega)$ be a nonnegative function. Set $\Omega_0 := \{x \in \Omega : a(x) = 0\}$ and consider the spaces

$$V := L^2([0, T], L^2(\Omega \backslash \Omega_0))$$

and

$$V_1 := \left\{ u \in L^2([0, T], W^{1,2}(\Omega)) : \int_0^T \int_\Omega a(x)|\partial_t u(s, x)|^2 \, dx ds < \infty \right\}.$$

Lemma 1.1 *The embedding $V_1 \subset V$ is compact for $n \leqslant 3$.*

Proof We set

$$\Omega_\delta^+ := \{x \in \Omega, \ a(x) > \delta\}, \quad \delta \geqslant 0.$$

Then, obviously, $\Omega_0^+ = \Omega \backslash \Omega_0$. Moreover, due to the continuity of the Lebesgue measure, we have

$$\lim_{\delta \to 0} \operatorname{mes}\{\Omega_0^+ \backslash \Omega_\delta^+\} = 0.$$

On the other hand, due to Hölder inequality and embedding $W^{1,2} \subset L^6$ (since $n \leqslant 3$), we have

$$\|\chi_X v\|_{L^2(\Omega)} \leqslant \operatorname{mes}\{X\}^{1/3} \|v\|_{L^6(X)} \leqslant C \operatorname{mes}\{X\}^{1/3} \|v\|_{W^{1,2}(\Omega)}$$

for any set $X \subset \Omega$ and, consequently,

$$\|\chi_X v\|_{L^2([0,T] \times \Omega)} \leqslant C \operatorname{mes}\{X\}^{1/3} \|v\|_{V_1}$$

where the constant C is independent of $v \in V_1$. Thus, for verifying the compactness of the embedding $V_1 \subset V$, it is sufficient to verify the compactness of the embedding

$$V_1 \subset V^\delta, \quad V^\delta := L^2([0, T] \times \Omega_\delta^+)$$

for any *positive* δ.

Let now $\delta > 0$ be fixed. Then, according to the Fréchet-Kolmogorov theorem, we need to verify that there exists a function $\mu : \mathbb{R}_+ \to \mathbb{R}_+, \ \lim_{z \to 0+} \mu(z) = 0$ such that

$$\int_0^T \int_\Omega \chi_{\Omega_\delta^+}(x)|u(x, t+s) - u(x, t)|^2 \, dx \, dt \leqslant \mu(|s|), \ s \in \mathbb{R} \tag{1.6}$$

and

$$\int_0^T \int_\Omega |\chi_{\Omega_\delta^+}(x+h)u(x+h,t) - \chi_{\Omega_\delta^+}(x)u(x,t)|^2 \, dx \, dt \leqslant \mu(|h|), \quad h \in \mathbb{R}^3 \tag{1.7}$$

uniformly with respect to all u belonging to the unit ball in V_1 (in these estimates function u is assumed to be extended by zero for $(x,t) \notin (0,T) \times \Omega$).

Let us first verify (1.6). Let $s > 0$ (the case $s < 0$ can be considered analogously). Then, using the obvious formula

$$u(x,t+s) - u(x,t) = s \int_0^1 \partial_t u(x,t+\kappa s) \, d\kappa$$

together with the fact that $a(x) > \delta$ if $x \in \Omega_\delta^+$, we have

$$\int_0^{T-s} \int_\Omega \chi_{\Omega_\delta^+}(x)|u(x,t+s) - u(x,t)|^2 \, dx \, dt$$

$$\leqslant s \int_0^T \int_\Omega \chi_{\Omega_\delta^+}(x)|\partial_t u(x,t)|^2 \, dx \, dt$$

$$\leqslant \delta^{-1} s \int_0^T \int_\Omega a(x)|\partial_t u(x,t)|^2 \, dx \, dt \leqslant \delta^{-1} s \|v\|_{V_1}^2. \tag{1.8}$$

On the other hand, using that

$$\|u(x,\cdot)\|_{L^\infty([0,T])} \leqslant C(\|\partial_t u(x,\cdot)\|_{L^2([0,T])} + \|u(x,\cdot)\|_{L^2([0,T])}),$$

we obtain

$$\int_{T-s}^T \int_\Omega \chi_{\Omega_\delta^+}(x)|u(x,t+s) - u(x,t)|^2 \, dx \, dt$$

$$\leqslant Cs \int_\Omega \chi_{\Omega_\delta^+}(x) \left(\|\partial_t u(x,\cdot)\|_{L^2([0,T])}^2 + \|u(x,\cdot)\|_{L^2([0,T])}^2 \right) dx$$

$$\leqslant C\delta^{-1} s \int_0^T \int_\Omega a(x)(|\partial_t u(x,t)|^2 + |u(x,t)|^2) \, dx \, dt \leqslant C\delta^{-1} \|u\|_{V_1}^2 \tag{1.9}$$

Estimates (1.8) and (1.9) show that (1.6) holds with $\mu(z) := 2C\delta^{-1}z$. Let us now verify (1.7). Indeed, due to the estimate

$$\left| \chi_{\Omega_\delta^+}(x+h)u(x+h,t) - \chi_{\Omega_\delta^+}(x)u(x,t) \right|$$

$$\leqslant \left| \chi_{\Omega_\delta^+}(x+h) - \chi_{\Omega_\delta^+}(x) \right| \cdot |u(x,t)| + |u(x+h,t) - u(x,t)|$$

and embedding $W^{1,2} \subset L^6$, we have

$$\int_0^T \int_\Omega \left| \chi_{\Omega_\delta^+}(x+h)u(x+h,t) - \chi_{\Omega_\delta^+}(x)u(x,t) \right|^2 dx\, dt$$

$$\leqslant CT\|u\|_{V_1}^2 \left(\int_\Omega |\chi_{\Omega_\delta^+}(x+h) - \chi_{\Omega_\delta^+}(x)|^3\, dx \right)^{2/3}$$

$$+ \int_0^T \int_\Omega |u(x+h,t) - u(x,t)|^2\, dx\, dt. \tag{1.10}$$

The first term on the right-hand side of (1.10) tends to zero since $\chi_{\Omega_\delta^+} \in L^\infty(\Omega) \subset L^3(\Omega)$ and the second one tends to zero uniformly with respect to u analogously to (1.6). Thus, estimates (1.6) and (1.7) are verified and Lemma 1.1 is proved.   □

1.2 Poincaré, Wirtinger, and Friedrichs Inequalities

A useful tool when dealing with Sobolev spaces and partial differential equations is the extension of classical Poincare inequality, such as the Poincare-Wirtinger (sometimes called Wirtinger) and Friedrichs inequalities. These type of inequalities usually provide Sobolev embeddings and compactness result as well as are of great importance in the modern, direct methods of the calculus variation. The goal of this section is to present these well-known results which we will use in the mathematical analysis of mitochondrial swelling scenarios in vivo, in vitro, and in silico. We start with the classical Poincare inequality for $W_0^{1,2}(\Omega)$ spaces.

Assume that Ω is a bounded open subset of the n-dimensional Euclidean space $\mathbb{R}^n$ with a Lipschitz domain (i.e., Ω is an open bounded Lipschitz domain). Then there exists a constant C depending only on Ω such that for every function u in the Sobolev space $W_0^{1,2}(\Omega)$ it holds that

$$\|u\|_{L^2(\Omega)} \leqslant C\|\nabla_x u\|_{L^2(\Omega)}.$$

The following proposition is called Wirtinger inequality.

Proposition 1.1 *Let* $u \in W^{1,2}(\Omega)$ *with zero mean value, that is,*

$$\int_\Omega u\, dx = 0.$$

Then there exists a constant $C_w > 0$, *such that the following estimate holds:*

$$\|u\|_{L^2(\Omega)} \leqslant C_w\|\nabla_x u\|_{L^2(\Omega)}. \tag{1.11}$$

A proof of these inequalities can be found in [2].

Corollary 1.1 *Let $u \in H^2(\Omega)$ satisfy the conditions of Proposition 1.1 and in addition assume that*

$$\|\nabla_x u\|^2_{L^2(\Omega)} = (-\Delta_x u, u)_{L^2(\Omega)}. \tag{1.12}$$

Then it holds with the same constants $C_w > 0$:

$$\|u\|_{L^2(\Omega)} \leqslant C_w \|\nabla_x u\|_{L^2(\Omega)} \leqslant C_w^2 \|\Delta_x u\|_{L^2(\Omega)}.$$

Proof Due to Cauchy-Schwarz and Wirtinger inequality it follows immediately

$$\|\nabla_x u\|^2_{L^2(\Omega)} = (-\Delta_x u, u)_{L^2(\Omega)} \leqslant \|\Delta_x u\|_{L^2(\Omega)} \|u\|_{L^2(\Omega)} \leqslant C_w \|\Delta_x u\|_{L^2(\Omega)} \|\nabla_x u\|_{L^2(\Omega)}$$

and, consequently,

$$\|\nabla_x u\|_{L^2(\Omega)} \leqslant C_w \|\Delta_x u\|_{L^2(\Omega)}.$$

Remark 1.1 Condition (1.12) is not very restrictive. If u satisfies homogeneous Dirichlet or Neumann boundary conditions, then it follows that

$$
\begin{aligned}
(-\Delta_x u, u)_{L^2(\Omega)} &= \int_\Omega -\Delta_x u \cdot u \, dx = -\int_{\partial\Omega} \nabla_x u \cdot u\vec{n} \, dS \\
&+ \int_\Omega \nabla_x u \cdot \nabla_x u \, dx = \int_\Omega \nabla_x u \cdot \nabla_x u \, dx \\
&= \|\nabla_x u\|^2_{L^2(\Omega)}.
\end{aligned}
$$

For the $W^{1,p}(\Omega)$ spaces the classical Poincare and Wirtinger inequality can be formulated as follows:

Proposition 1.2 *Let Ω be a bounded open subset of $\mathbb{R}^n$ with a Lipschitz boundary and $p \in [1, \infty]$. Then there exists a constant C depending only on Ω and p, such that for every function $u \in W^{1,p}(\Omega)$ holds:*

$$\|u - \langle u \rangle_\Omega\|_{L^p(\Omega)} \leqslant C \|\nabla_x u\|_{L^p(\Omega)},$$

where

$$\langle u \rangle_\Omega := \frac{1}{|\Omega|} \int_\Omega u(y) \, dy$$

is the average value of u over Ω, with $|\Omega|$ standing for the Lebesgue measure of the domain Ω.

The following inequality, which is called Friedrichs inequality, is a generalization of Poincare's inequality for the functions satisfying arbitrary boundary conditions that we will use in the study of mitochondrial swelling scenario with Robin boundary conditions.

Let a be a given real valued function satisfying

(a) $0 \leqslant a \leqslant M$, where M is some positive number and
(b) $a \neq 0$ in Ω, where Ω is a bounded domain in $\mathbb{R}^n$.

We denote by

$$\phi(u) := \left(\int_{\partial\Omega} a(x)|u(x)|^2 \, dx \right)^{1/2} \tag{1.13}$$

and we will consider $\phi : H^1(\Omega) \to [0, \infty)$. Then the following assertion holds.

Lemma 1.2 *Let a be a given function satisfying conditions a) and b) above and $\phi : H^1(\Omega) \to [0, \infty)$ be the functional given by (1.13). Then there exists a constant $C > 0$ such that*

$$\|u\|_{L^2(\Omega)} \leqslant C(\|\nabla_x u\|_{L^2(\Omega)} + \phi(u)). \tag{1.14}$$

Proof Assume the contrary. Then for any $n \in \mathbb{N}$, there exists $u_n \in H^1(\Omega)$ such that

$$\|u_n\|_{L^2(\Omega)} \geqslant n(\|\nabla_x u_n\|_{L^2(\Omega)} + \phi(u_n))$$

or equivalently

$$\|u_n\|_{L^2(\Omega)} \geqslant n\varphi(u_n), \tag{1.15}$$

where

$$\varphi(u_n) := \|\nabla_x u_n\|_{L^2(\Omega)} + \phi(u_n).$$

We set $\tilde{u}_n := \frac{u_n}{\varphi(u_n)}$. Then it follows from (1.15) that $\|\tilde{u}_n\|_{L^2(\Omega)} \geqslant n$ and $\tilde{u}_n \in H^1(\Omega)$. Therefore,

$$\|\tilde{u}_n\|_{L^2(\Omega)} \to \infty \quad \text{and} \quad \|\nabla_x \tilde{u}_n\|_{L^2(\Omega)} + \phi(\tilde{u}_n) = 1.$$

Let $v_n = \frac{\tilde{u}_n}{\|\tilde{u}_n\|_{L^2(\Omega)}}$. Then we obtain

$$\varphi(v_n) = \frac{1}{\|\tilde{u}_n\|_{L^2(\Omega)}} \left(\|\nabla_x \tilde{u}_n\|_{L^2(\Omega)} + \phi(\tilde{u}_n) \right) = \frac{1}{\|\tilde{u}_n\|_{L^2(\Omega)}}.$$

Hence $\varphi(v_n) \to 0$ as $n \to \infty$. Since

$$\phi(v_n) \leqslant \varphi(v_n) \quad \text{and} \quad \|\nabla_x v_n\|_{L^2(\Omega)} \leqslant \varphi(v_n), \tag{1.16}$$

we have

$$\phi(v_n) \to 0 \quad \text{and} \quad \|\nabla_x v_n\|_{L^2(\Omega)} \to 0 \text{ as } n \to \infty.$$

Taking into account that $\|v_n\|_{L^2(\Omega)} = 1$, then from (1.16) it follows that $v_n \in H^1(\Omega)$ and $\|v_n\|_{H^1(\Omega)} \leqslant C_*$, where C_* does not depend on n. Hence there exists a subsequence of $\{v_n\}$, say $\{v_{n_k}\}$, such that

$$\nabla_x v_{n_k} \xrightarrow{w} \nabla_x v_\infty \quad \text{weakly in } H^1(\Omega)$$

$$v_{n_k} \xrightarrow{s} v_\infty \quad \text{strongly in } L^2(\Omega),$$

where $v_\infty \in H^1(\Omega)$. Since $\varphi(v_n) \to 0$ as $n \to \infty$, it follows that

$$0 = \lim_{n\to\infty} \varphi(v_n) = \lim_{k\to\infty} \varphi(v_{n_k}) = \lim_{k\to\infty} \inf(\|\nabla_x v_{n_k}\|_{L^2(\Omega)} + \phi(v_{n_k}))$$

$$\geqslant \|\nabla_x v_\infty\|_{L^2(\Omega)} + \phi(v_\infty). \tag{1.17}$$

In (1.17) we used the fact that the functional $\phi : H^1(\Omega) \to [0, \infty)$ is weakly sequentially continuous. Thus we have

$$0 \geqslant \|\nabla_x v_\infty\|_{L^2(\Omega)} + \phi(v_\infty) = 0$$

and consequently both $\|\nabla_x v_\infty\|_{L^2(\Omega)} = 0$ and $\phi(v_\infty) = 0$. Next we would like to show that $v_\infty \equiv C_\infty$, where C_∞ is constant and this constant is identically zero. Assume that $v_\infty \not\equiv C_\infty$. Then we can decompose $v_\infty \in H^1(\Omega)$ into

$$v_\infty = C_\infty + v_\infty^\perp,$$

where $v_\infty^\perp \in H_\perp = \{v \in L^2(\Omega) \mid \int_\Omega v \, dx = 0\}$. Using Wirtinger's inequality (1.11) we deduce that $v_\infty^\perp \equiv 0$ almost everywhere in Ω, and consequently $v_\infty \equiv C_\infty$ almost everywhere in Ω. To show that $C_\infty = 0$ we assume the contrary, that is, $C_\infty \neq 0$. Then on one hand

$$\phi(1) = \phi\left(\frac{1}{C_\infty} \cdot C_\infty\right) = \frac{1}{|C_\infty|}\phi(C_\infty) = 0$$

and on the other hand $\phi(1) = (\int_{\partial\Omega} a(x) \, dx)^{1/2} > 0$. This contradiction leads to (1.14). Lemma 1.2 is proved.   □

Remark 1.2 Note that the assertion of Lemma 1.2 remains valid if we replace the functional $\phi : H^1(\Omega) \to [0, \infty)$ in (1.13) by a more general one $B : H^1(\Omega) \to [0, \infty)$ satisfying

(i) $B(\lambda u) \leqslant |\lambda| B(u) \quad \forall \lambda \in \mathbb{R}$,
(ii) B is sequentially weakly continuous,
(iii) $B(1) = c_0 > 0$.

1.3 Nemytski Operators

The investigation of nonlinear equations in the following chapters relies on properties of mappings of the form $u \mapsto f(u)$ in the spaces $C^\alpha(\bar{\Omega})$ and $L^p(\Omega)$, $H^l(\Omega)$.

Definition 1.1 Let $\Omega \subset \mathbb{R}^n$ be a domain. We say that a function

$$\Omega \times \mathbb{R}^m \ni (x, u) \longmapsto f(x, u) \in \mathbb{R}$$

satisfies the Caratheodory conditions if

$$u \longmapsto f(x, u) \quad \text{is continuous for almost every } x \in \Omega$$

and

$$x \longmapsto f(x, u) \quad \text{is measurable for every } u \in \Omega.$$

Given any f satisfying the Caratheodory conditions and a function $u : \Omega \to \mathbb{R}^m$, we can define another function by composition

$$F(u)(x) := f(x, u(x)). \tag{1.18}$$

The composed operator F is called a Nemytskii operator. In this section we state some important results on composition in spaces $C^\alpha(\bar{\Omega})$, $L^p(\Omega)$, $H^l(\Omega)$ (some of them without proofs, they can be found in [5, 11]).

Proposition 1.3 *Let* $\Omega \subset \mathbb{R}^n$ *be a bounded domain and*

$$\Omega \times \mathbb{R}^m \ni (x, u) \longmapsto f(x, u) \in \mathbb{R}$$

satisfy the Caratheodory conditions. In addition, let

$$|f(x, u)| \leqslant f_0(x) + c(1 + |u|)^r, \tag{1.19}$$

where $f_0 \in L^{p_0}(\Omega)$, $p_0 \geqslant 1$, and $rp_0 \leqslant p_1$. Then the Nemytskii operator F defined by (1.18) is bounded from $L^{p_1}(\Omega)$ into $L^{p_0}(\Omega)$ and

$$\|F(u)\|_{0,p_0} \leqslant C_1 \left(1 + \|u\|_{p_1}^r\right) \tag{1.20}$$

Proof By (1.19) and (1.2)

$$\|\cdot\|_{o,p_0} \leqslant \|f_0(x)\|_{o,p_0} + C\|1\|_{o,p_0} + C\||u|^r\|_{o,p_0}$$

$$\leqslant C' + C \left(\int_\Omega |u|^{rp_0} dx\right)^{\frac{1}{p_0}}$$

$$= C' + \|u\|_{0,p_0 r}^r. \tag{1.21}$$

Since Ω is bounded, then by Hölder inequality

$$\|v\|_{0,q} \leqslant C(\Omega)\|v\|_{0,p} \text{ when } 1 \leqslant q \leqslant p, \ v \in L^p(\Omega), \tag{1.22}$$

where $C(\Omega) = (\text{mes}(\Omega))^{\frac{1}{q} - \frac{1}{p}}$. Inequalities (1.21) and (1.22) with $q = rp_0$ and $p = p_1$ imply (1.20).   □

It is well known that the notions of continuity and boundedness of a nonlinear operator are independent of one another [5]. It turns out that the following is valid.

Theorem 1.3 *Let $\Omega \subset \mathbb{R}^n$ be a bounded domain and let*

$$\Omega \times \mathbb{R}^m \ni (x, u) \longmapsto f(x, u) \in \mathbb{R}$$

satisfy the Caratheodory conditions. In addition, let $p \in (1, \infty)$ and $g \in L^q(\Omega)$ (where $\frac{1}{p} + \frac{1}{q} = 1$) be given, and let f satisfy

$$|f(x, u)| \leqslant C|u|^{p-1} + g(x).$$

Then the Nemytskii operator F defined by (1.18) is a bounded and continuous map from $L^p(\Omega)$ to $L^q(\Omega)$.

For a more detailed treatment, the reader could consult [5, 11].

Theorem 1.4 *Let Ω be a bounded domain in $\mathbb{R}^n$ with smooth boundary and let*

$$\Omega \times \mathbb{R} \ni (x, u) \mapsto f(x, u) \in \mathbb{R}$$

satisfy the Caratheodory conditions. Then f induces

1) a continuous mapping from $H^s(\Omega)$ into $H^s(\Omega)$ if $f \in C^s$,
2) a continuously differentiable mapping from $H^s(\Omega)$ into $H^s(\Omega)$ if $f \in C^{s+1}$, where in both cases $s > n/2$, $s \in \mathbb{N}$.

Proof First we consider the simplest case, that is $f = f(u)$ is independent of x. By the Sobolev embedding theorem, we have $H^s(\Omega) \subset C(\bar{\Omega})$. Hence we have $f(u) \in C(\bar{\Omega})$ for every $u \in H^s(\Omega)$. Moreover, if u is in $C^s(\bar{\Omega})$, we can obtain the derivatives of $f(u)$ by the chain rule, and in the general case, we can use approximation by smooth functions. Note that all derivatives of $f(u)$ have the form of a product involving a derivative of f and derivatives of u. The first factor is in $C(\bar{\Omega})$, while any l-th derivative of u lies in $H^{s-l}(\Omega)$, which embeds into $L^{2n/(n-2(s-l))}(\Omega)$ if $s - l < \frac{n}{2}$. We can use this fact and Hölder inequality to show that all derivatives of $f(u)$ up to order s are in $L^2(\Omega)$; moreover, it is clear from this argument that f is actually continuous from $H^s(\Omega)$ into $H^s(\Omega)$. A proof of the differentiability in this special case is that $f = f(u)$ is based on the relation

$$f(u) - f(v) = \int_0^1 f_u'(v + \theta(u - v))(u - v)d\theta$$

and the same arguments as before.

Let us now consider the general case, that is $f = f(x, u)$. Let $|\alpha| \leq s$. We must show that

$$u \longmapsto D^\alpha F(u) \tag{1.23}$$

defines a continuous map of $H^s(\Omega)$ into $L^2(\Omega)$. It is not difficult to see that (1.23) is the finite sum of operators of the form

$$u \longmapsto g(\cdot, u) \cdot D^\gamma u,$$

where $|\gamma| = \gamma_1 + \cdots + \gamma_n \leq s$, while g is a partial derivative of f order at most s. It is obvious that D^γ is continuous from $H^s(\Omega)$ into $L^2(\Omega)$ for $|\gamma| \leq s$. On the other hand, the continuous embedding of $H^s(\Omega)$ in $C(\bar{\Omega})$ implies that

$$u \longmapsto g(\cdot, u)$$

is continuous from $H^s(\Omega)$ into $C(\bar{\Omega})$. Thus

$$u \longmapsto g(\cdot, u) \cdot D^\gamma u$$

defines a continuous map of $H^s(\Omega)$ into $L^2(\Omega)$ and hence so does $u \longmapsto D^\alpha F(u)$.
$\square$

For $p \in \mathbb{N}$, let $\tilde{p}$ be the number of multiindices α with $|\alpha| \leq p$.

Corollary 1.2 *An analogous result is valid for a continuity of the operator*

$$F(u) = f(\cdot, u, \ldots, D^p u) : H^{s+p}(\Omega) \to H^s(\Omega),$$

where $p, s \in \mathbb{N}$ with $s > \frac{n}{2}$ and $f : \Omega \times \mathbb{R}^{\tilde{p}} \to \mathbb{R}$ is C^s.

Corollary 1.3 *Let $p, s \in \mathbb{N}$ with $s > \frac{n}{2}$ and*

$$f : \Omega \times \mathbb{R}^{\tilde{p}} \to \mathbb{R} \text{ be } C^{s+1}.$$

Then the operator $F : H^{s+p}(\Omega) \to H^s(\Omega)$ defined by

$$F(u)(x) = f(x, u(x), \ldots, D^p u(x))$$

is Frechet differentiable from $H^{s+p}(\Omega)$ into $H^s(\Omega)$.

We have the following continuity and C^1-differentiability results for a nonlinear differential operator of the form $A(u)(x) = f(x, u(x), \ldots D^{2p} u(x))$ in the Hölder spaces. They are based on Theorems 1.5 and 1.6.

Let $p \in \mathbb{N}$ and $\tilde{p}$ denote as before the number of multiindices with $|\alpha| \leqslant p$. Let Ω be a bounded domain in $\mathbb{R}^n$.

Theorem 1.5 *Let the function $f(x, y) = f(x, y_1, \ldots, y_{\tilde{p}})$ be defined on $\bar{\Omega} \times \mathbb{R}^{\tilde{p}}$ which satisfies the following conditions:*

1) $f(x, 0) = 0$
2) For any $R > 0$, $\displaystyle\sup_{|y| \leqslant R} \left| \frac{\partial^2 f}{\partial y_i \partial y_j}(\cdot, y) \right| \leqslant C(R)$, $\displaystyle\sup_{|y| \leqslant R} \|f(\cdot, y)\|_{C^{1, \alpha}(\bar{\Omega})} \leqslant C(R)$,
where $C(R)$ is a constant depending on R.

Let $u_1, \ldots, u_{\tilde{p}} \in C^\alpha(\bar{\Omega})$, $0 < \alpha < 1$, $\|u_i\|_{C^\alpha(\bar{\Omega})} \leqslant R$, $i = 1, \ldots, \tilde{p}$. Then

$$\|f(\cdot, u_1, \ldots, u_{\tilde{p}})\|_{C^\alpha(\bar{\Omega})} \leqslant C_1(R) \sum_{i=1}^{\tilde{p}} \|u_i\|_{C^\alpha(\bar{\Omega})}. \tag{1.24}$$

Proof Obviously,

$$f(x, y, \ldots, y_{\tilde{p}}) = \int_0^1 \frac{d}{dt} f(x, ty_1, \ldots, ty_{\tilde{p}}) dt$$

$$= \sum_{j=1}^{\tilde{p}} y_j \int_0^1 \frac{\partial f(x, ty_1, \ldots, ty_{\tilde{p}})}{\partial y_j} dt$$

$$= \sum_{j=1}^{\tilde{p}} \varphi_j(x, y_1, \ldots, y_{\tilde{p}}) y_j,$$

where

$$\varphi_j(x, y_1, \ldots, y_{\tilde{p}}) = \int_0^1 \frac{\partial f(x, ty_1, \ldots, ty_{\tilde{p}})}{\partial y_j} dt.$$

Hence

$$f(x, u_1(x), \ldots, u_{\tilde{p}}(x)) = \sum_{j=1}^{\tilde{p}} \varphi_j(x, u_1(x), \ldots, u_{\tilde{p}}(x)) u_j(x).$$

Since $C^\alpha(\bar{\Omega}), 0 < \alpha < 1$ is a Banach algebra, we have

$$\|f(\cdot, u_1, \ldots u_{\tilde{p}})\|_{C^\alpha} \leqslant \sum_{j=1}^{\tilde{p}} \|\varphi_j(\cdot, u_1, \ldots u_{\tilde{p}})\|_{C^\alpha} \|u_j\|_{C^\alpha}.$$

Hence we have to prove that

$$\sup_{|y| \leqslant R} \|\varphi_j(\cdot, u_1, \ldots, u_{\tilde{p}})\|_{C^\alpha} < C_1(R).$$

Indeed

$$|\varphi_j(x + \xi, u_1(x + \xi), \ldots, u_{\tilde{p}}(x + \xi)) - \varphi_j(x, u_1(x), \ldots, u_{\tilde{p}}(x))|$$
$$\leqslant |\varphi_j(x + \xi, u_1(x + \xi), \ldots, u_{\tilde{p}}(x + \xi)) - \varphi_j(x, u_1(x + \xi), \ldots, u_{\tilde{p}}(x + \xi))|$$
$$+ |\varphi_j(x, u_1(x + \xi), \ldots, u_{\tilde{p}}(x + \xi)) - \varphi_j(x, u_1(x), \ldots, u_{\tilde{p}}(x))|. \quad (1.25)$$

The first term on the right-hand side of (1.25) is bounded by $C(R)|\xi|^\alpha$. The second term is bounded by

$$\sup_{|y| \leqslant R} \left| \frac{\partial \varphi_j}{\partial y_k} \right| |\varphi_j(x, u_1(x + \xi), \ldots,$$

$$u_{\tilde{p}}(x + \xi)) - \varphi_j(x, u_1(x), \ldots, u_{\tilde{p}}(x))| \leqslant C(R) R |\xi|^\alpha. \quad (1.26)$$

Estimates (1.25) and (1.26) yield (1.24). □

Theorem 1.6 *Let the function* $f(x, y) = f(x, y_1, \ldots, y_{\tilde{p}})$ *be defined on* $\bar{\Omega} \times R^{\tilde{p}}$ *satisfy the following conditions:*

1) $f(\cdot, 0) \equiv 0, \ \nabla_y f(\cdot, 0) \equiv 0$

2) For any $R > 0, \ \sup_{|y| \leqslant R} \|f(\cdot, y)\|_{C^{2,\alpha}(\bar{\Omega})} \leqslant C(R)$ *and* $\sup_{|y| \leqslant R} \left| \frac{\partial^3 f}{\partial y_i \partial y_j \partial y_k}(\cdot, y) \right| \leqslant C(R),$

where $C(R)$ *is a constant depending on* R. *Let as before,* $u_1(x), \ldots, u_{\tilde{p}}(x) \in C^\alpha(\bar{\Omega})$ *with* $\|u_i\|_{C^\alpha(\bar{\Omega})} \leqslant R, i = 1, \ldots, \tilde{p}$. *Then the following estimate holds.*

$$\|f(\cdot, u_1, \ldots, u_{\tilde{p}})\|_{C^\alpha(\bar{\Omega})} \leqslant C_2(R) \sum_{i=1}^{\tilde{p}} \|u_i\|_{C^\alpha}^2. \quad (1.27)$$

Proof Obviously we have

$$f(x, y_1, \ldots, u_{\tilde{p}}) = \sum_{i,j=1}^{\tilde{p}} g_{ij}(x, y_1, \ldots, y_{\tilde{p}}) y_i y_j,$$

so we can write

$$f(x, u_1(x), \ldots, u_{\tilde{p}}(x)) = \sum_{i,j=1}^{\tilde{p}} g_{ij}(x, u_1(x), \ldots, u_{\tilde{p}}(x)) u_i(x) u_j(x)$$

and we have

$$\|f(\cdot, u_1, \ldots, u_{\tilde{p}})\|_{C^\alpha(\bar{\Omega})}$$

$$\leqslant \sum_{i,j=1}^{\tilde{p}} \|g_{ij}(\cdot, u_1, \ldots, u_{\tilde{p}})\|_{C^\alpha(\bar{\Omega})} \|u_i\|_{C^\alpha} \|u_j\|_{C^\alpha}. \tag{1.28}$$

Due to Theorem 1.5 we obtain

$$\|g_{ij}(\cdot, u_1, \ldots, u_{\tilde{p}})\|_{C^\alpha(\bar{\Omega})} \leqslant C_0(R). \tag{1.29}$$

Hence the estimates (1.28) and (1.29) yield (1.27)

$$\|f(\cdot, u_1, \ldots, u_{\tilde{p}})\|_{C^\alpha(\bar{\Omega})} \leqslant C_2(R) \sum_{i=1}^{\tilde{p}} \|u_i\|_{C^\alpha}^2.$$

We apply Theorems 1.5 and 1.6 to the operator

$$A(u)(x) = f(x, u(x), \ldots, D^{2p} u(x)),$$

where function $f(x, y_1, \ldots, y_{\tilde{p}})$ satisfies conditions of Theorems 1.5 and 1.6, respectively. Hence we have

$$\|A(u)\|_{C^{2p,\alpha}} \leqslant C(R) \|u\|_{C^\alpha}.$$

Moreover, it follows from Theorem 1.6 that $A \in C^1$, $A'(0) = 0$, and

$$\|A'(u+h) - A'(u)\|_{L(C^{2p,\alpha}, C^\alpha)} \leqslant C \|h\|_{C^{2p,\alpha}(\bar{\Omega})}.$$

$\square$

Remark 1.3 Continuity and differentiability of the operator $A(u)(x) = f(x, u(x), \ldots, D^{2p} u(x))$ between $C^{2p,\alpha}(\bar{\Omega})$ and $C^\alpha(\bar{\Omega})$ remain valid under slightly weaker conditions on a given function $f(x, y_1, \ldots, y_{\bar{p}})$. We leave these as exercises for the reader.

The following lemma on the smoothness relations between u and $f(u)$ plays a decisive role in the study of life science problems (see [4]).

Lemma 1.3 *Let the function* $f \in C^2(\mathbb{R}, \mathbb{R})$ *satisfies*

$$C_1|u|^{p-1} \leqslant f'(u) \leqslant C_1|u|^{p-1},$$

$p > 1$, *with* C_1 *and* C_2 *some positive constants. Then, for every* $s \in (0, 1)$ *and* $1 < q \leqslant \infty$, *we have*

$$\|u\|_{W^{s/p,pq}(\Omega)} \leqslant C_p \|f(u)\|_{W^{s,q}(\Omega)}^{1/p}$$

where the constant C_p *is independent of* u.

Proof Indeed, let f^{-1} be the inverse function to f. Then, due to conditions on f, the function $G(v) := \text{sgn}(v)|f^{-1}(v)|^p$ is nondegenerate and satisfies

$$C_2 \leqslant G'(v) \leqslant C_1,$$

for some positive constants C_1 and C_2. Therefore, we have

$$|f^{-1}(v_1) - f^{-1}(v_2)|^p \leqslant C_p|G(v_1) - G(v_2)| \leqslant C_p'|v_1 - v_2|,$$

for all $v_1, v_2 \in \mathbb{R}$. Finally, according to the definition of the fractional Sobolev spaces (see, e.g., [8, 9]),

$$\|f^{-1}(v)\|_{W^{s/p,qp}(\Omega)}^{pq} := \|f^{-1}(v)\|_{L^{pq}(\Omega)}^{pq} + \int_\Omega \int_\Omega \frac{|f^{-1}(v(x)) - f^{-1}(v(y))|^{pq}}{|x - y|^{n+sq}} \, dx \, dy$$

$$\leqslant C\|v\|_{L^q(\Omega)}^q + C_p' \int_\Omega \int_\Omega \frac{|v(x) - v(y)|^q}{|x - y|^{n+sq}} \, dx \, dy$$

$$= C_p''\|v\|_{W^{s,q}(\Omega)}^q,$$

where we have implicitly used that $f^{-1}(v) \sim \text{sgn}(v)|v|^{1/p}$. Lemma 1.3 is proved.

$\square$

1.4 Subdifferential

In the study of well posedness of solutions of mitochondria swelling models in vitro and in silico (see Chaps. 6–7) we deal with evolution equations involving subdifferentials. For the convenience of the reader in this section we briefly recall

Fig. 1.1 A smooth function

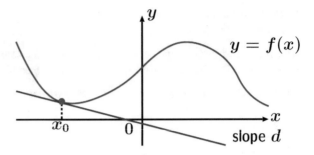

Fig. 1.2 A non-smooth
function $f(x) = |x|$

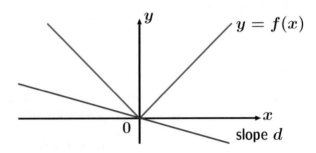

some basic facts related to subdifferentials. We start with the scalar case. It is well
known that for a smooth function $f = f(x)$ (Fig. 1.1):

- the differential coefficient d_{x_0} of f at x_0 is the slope of the tangent to the graph
  of $y = f(x)$ at the point x_0;
- the derivative $f' = f'(x)$ of f is the function mapping x_0 to the differential
  coefficient d_{x_0}.

As for a non-smooth function $f(x)$, the slope of the tangential to the graph
of $f(x)$ may NOT be uniquely determined (Fig. 1.2). For example, the function
$f(x) = |x|$ is not differentiable at $x = 0$. The notion of subdifferential was
introduced as a generalization of derivative for non-smooth convex functions.

Definition 1.2 Subdifferential ∂f of a convex function f at a point x_0 is a set
defined by

$$\partial f(x_0) = \left\{ \text{the slope of 'possible' tangents to } y = f(x) \text{ at } x_0 \right\}.$$

If f is smooth, then $\partial f(x) = \{f'(x)\}$ (Fig. 1.3).

Example 1.1 $f(x) = |x|$.

$$\partial f(x_0) = 1 \text{ if } x_0 > 0, \quad \partial f(0) = [-1, 1], \quad \partial f(x_0) = -1 \text{ if } x_0 < 0.$$

Fig. 1.3 A non-smooth function

$$f(x) = \begin{cases} 0 & \text{if } x \leqslant 0, \\ \infty & \text{if } x > 0. \end{cases}$$

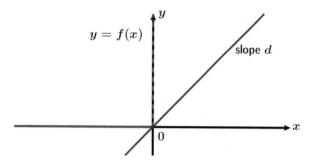

Fig. 1.4 Monotonicity

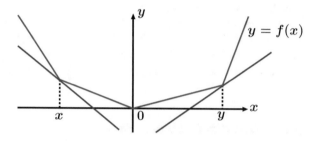

Example 1.2 $f(x) = \begin{cases} 0 & \text{if } x \leqslant 0, \\ \infty & \text{if } x > 0. \end{cases}$

$\partial f(x_0) = \{0\}$ if $x_0 < 0$, $\quad \partial f(0) = [0, \infty)$, $\quad \partial f(x_0) = \varnothing$ if $x_0 > 0$.

Monotonicity of Subdifferentials

∂f is monotonely increasing ($\Leftrightarrow$ the convexity of f), i.e.,

$$x \leqslant y \quad \Rightarrow \quad \max \partial f(x) \leqslant \min \partial f(y).$$

Equivalently, $(\xi - \eta)(x - y) \geqslant 0$ for all $\xi \in \partial f(x), \eta \in \partial f(y)$ (Fig. 1.4).

Subdifferentials and PDEs

The notion of subdifferentials is extended to (non-smooth convex) function(al)s defined in infinite-dimensional spaces.

For example, the Laplace operator

$$-\Delta_x u(x, y, z) := -\left(\frac{\partial^2 u}{\partial x^2} + \frac{\partial^2 u}{\partial y^2} + \frac{\partial^2 u}{\partial z^2}\right)$$

can be expressed as a subdifferential operator $u \mapsto \partial f(u)$ in $L^2(\Omega)$ of

$$f(u) = \frac{1}{2} \iiint_{\Omega} |\nabla_x u(x, y, z)|^2 \, dx \, dy \, dz,$$

which is non-smooth but convex in $L^2(\Omega)$.

It is worth noting that various important PDEs, e.g., heat equation, porous medium equation, Navier-Stokes equation, are formulated as evolution equations (ODEs in infinite-dimensional spaces) involving subdifferentials. There the gradient structure and (maximal) monotonicity of ∂f play crucial roles.

Subdifferential in Hilbert Spaces

Definition 1.3 Let $(H, (\cdot, \cdot)_H)$ be a Hilbert space and let $\varphi : H \to (-\infty, \infty]$ be convex and lower semicontinuous with

$$D(\varphi) := \{u \in H : \varphi(u) < \infty\} \neq \emptyset.$$

Then the subdifferential operator $\partial \varphi : H \to 2^H$ is defined by

$$\partial \varphi(u) := \{\xi \in H : \varphi(v) - \varphi(u) \geqslant (\xi, v - u)_H \ \forall v \in H\}$$

for $u \in D(\varphi)$ with domain $D(\partial \varphi) := \{u \in D(\varphi) : \partial \varphi(u) \neq \emptyset\}$.

Remark 1.4 The notion of subdifferentials can be extended to topological linear spaces.

The following theorem contains several most important properties of subdifferentials in Hilbert spaces

Theorem 1.7

1. *$\partial \varphi$ is maximal monotone;*
   - *$(\xi - \eta, u - v)_H \geqslant 0$ for all $\xi \in \partial \varphi(u)$ and $\eta \in \partial \varphi(v)$;*
   - *For any $f \in H$ there exists $u \in D(\partial \varphi)$ such that*

$$u + \partial \varphi(u) \ni f.$$

2. *In particular, $\partial\varphi$ is demiclosed: If a sequence $\xi_n \in \partial\varphi(u_n)$ satisfies*

$$u_n \to u \quad \text{weakly in } H, \quad \xi_n \to \xi \quad \text{weakly in } H, \quad \limsup_{n\to\infty}(\xi_n, u_n)_H \leqslant (\xi, u)_H,$$

then $u \in D(\partial\varphi)$ and $\xi \in \partial\varphi(u)$.

3. *Chain rule for $\partial\varphi$:*
    If $u \in W^{1,2}(0, T; H)$ and there exists $\xi \in L^2(0, T; H)$ such that

$$\xi(t) \in \partial\varphi(u(t)) \quad \text{for a.e. } t \in (0, T)$$

then $\varphi(u)$ belongs to $W^{1,1}(0, T)$ and

$$\frac{d}{dt}\varphi(u(t)) = \left(g, \frac{du}{dt}(t)\right)_H \quad \text{for all } g \in \partial\varphi(u(t))$$

for a.e. $t \in (0, T)$.

4. *Fenchel-Moreau identity: The convex conjugate $\varphi^* : H \to (-\infty, \infty]$ of φ is defined by*

$$\varphi^*(f) := \sup_{v\in H} \left[(f, v)_H - \varphi(v)\right] \quad \text{for } f \in H.$$

Then $\varphi^{**} = \varphi$, $\partial\varphi^* = (\partial\varphi)^{-1}$ and

$$\varphi(u) + \varphi^*(f) = (f, u)_H \quad \text{iff} \quad f \in \partial\varphi(u).$$

Now we present several examples of subdifferentials in Hilbert spaces

Example 1.3 Let $H = L^2(\Omega)$ and define

$$\psi : H \to [0, \infty]$$

$$\varphi(u) := \begin{cases} \frac{1}{2}\int_\Omega |\nabla_x u(x)|^2\, dx & \text{if } u \in H_0^1(\Omega), \\ \infty & \text{otherwise.} \end{cases}$$

Then $\partial\varphi(u)$ coincides with the Dirichlet Laplacian, more precisely, for each $u \in D(\partial\varphi) = H^2(\Omega) \cap H_0^1(\Omega)$,

$$(\partial\varphi(u), v)_H = \int_\Omega \nabla_x u \cdot \nabla_x v\, dx \quad \text{for } v \in H_0^1(\Omega).$$

From the properties of subdifferentials, we have

- The Dirichlet Laplacian $-\Delta_x$ is maximal monotone in $L^2(\Omega)$.
- The chain-rule holds: $(-\Delta_x u(t), \partial_t u(t))_{L^2(\Omega)} = \frac{d}{dt}\varphi(u(t))$.

Example 1.4 Let α be a maximal monotone graph in $\mathbb{R}$ and define

$$\psi : H \to [0, \infty]$$

$$\psi(u) := \begin{cases} \int_\Omega \hat\alpha(u(x))\, dx & \text{if } \hat\alpha(u) \in L^1(\Omega), \\ \infty & \text{otherwise,} \end{cases}$$

where $\hat\alpha$ is the primitive function of α. Then $\partial\psi(u)$ coincides with $\alpha(u)$. From the properties of subdifferentials, we have

- $u \mapsto \alpha(u)$ is maximal monotone in $L^2(\Omega)$.
- The chain-rule holds: $(\alpha(u(t)), \partial_t u(t))_{L^2(\Omega)} = \frac{d}{dt}\psi(u(t))$.

Proposition 1.4 *Let $m > 2$. Function $\varphi : H \to [0, \infty]$ defined by*

$$\varphi(u) := \begin{cases} \frac{1}{m}\|u\|_m^m & \text{if } u \in L^m(\Omega) \\ \infty & \text{otherwise} \end{cases}$$

possesses in H the maximal monotone subdifferential

$$\partial\varphi(u) = -\Delta_x(|u|^{m-2}u). \tag{1.30}$$

Proof

1) *φ is proper*

   We have $\varphi(t) < \infty$ due to the embedding

   $$L^m(\Omega) \hookrightarrow L^2(\Omega) \subset\subset H^{-1}(\Omega) = H.$$

2) *φ is convex*

   In order to show the convexity, we define $\tilde\varphi := g \circ f$

   $$\tilde\varphi : L^m(\Omega) \to [0, \infty)$$

   with $f : L^m(\Omega) \to [0, \infty)$ and $g : [0, \infty) \to [0, \infty)$ given by

   $$f(u) := \|u\|_m \quad \text{and} \quad g(s) := \frac{1}{m}s^m.$$

Then f is convex by the triangle inequality valid of every norm and g is convex since $g''(s) \geqslant 0$ for every $s \in [0, \infty)$. Furthermore g is nondecreasing and it follows for every $\lambda \in (0, 1)$

$$g\left(f(\lambda u + (1 - \lambda)v)\right) \leqslant g\left(\lambda f(u) + (1 - \lambda)f(v)\right) \leqslant \lambda g\left(f(u)\right) + (1 - \lambda)g\left(f(v)\right),$$

thus $\tilde\varphi$ is convex. Knowing that, it immediately follows that

$$\varphi(u) := \begin{cases} \tilde{\varphi}(u) & \text{if } u \in L^m(\Omega) \\ \infty & \text{otherwise} \end{cases} \tag{1.31}$$

is also convex.

3) φ *is lower semicontinuous*

Here we again use the composition $\tilde{\varphi} = g \circ f$ as described above. It is a basic knowledge that every norm is lower semicontinuous with respect to weak convergence and so we have lower semicontinuity of f. Furthermore g as a polynomial is continuous and increasing on its domain $[0, \infty)$. This implies that $\tilde{\varphi}$ is lower semicontinuous as it is, e.g., proved in [3]. Again, this is also satisfied by the extension (1.31).

By Theorem 1.7 it follows that the subdifferential $\partial\varphi$ is maximal monotone, but it remains to show that it is given by (1.30). For that we will show that φ is Gâteaux differentiable and consequently $\partial\varphi(u) = \{u^*\}$. Then will prove that $u^* = -\Delta_x |u|^{m-2} u$ in $H^{-1}(\Omega)$.

4) φ *is Gâteaux differentiable*

For $h > 0$ we have to show

$$\lim_{h \to 0} \frac{\varphi(u + hv) - \varphi(u)}{h} = D\varphi(u)(v)$$

where the derivative $D\varphi(u)(v)$ of φ at u in direction v is linear and bounded in v.

In accordance with Definition 1.3, the subdifferential of φ is only defined for $u \in L^m(\Omega)$, hence we only need Gâteaux differentiability in $L^m(\Omega)$. Here we have $\varphi(u) = \frac{1}{m}\|u\|_m^m$ and it is, e.g., proven in [10], that this functional is Gâteaux (and also Fréchet) differentiable for $1 < m < \infty$. The proof is done using the auxiliary function $\psi(h) := \varphi(u + hv)$, which is differentiable in h. Then by the standard differentiability it holds

$$\psi'(0) = \lim_{h \to 0} \frac{\psi(h) - \psi(0)}{h} = \lim_{h \to 0} \frac{\varphi(u + hv) - \varphi(u)}{h}.$$

Hence it follows that $\partial\varphi(u)$ consists of only one element for every $u \in L^m(\Omega)$. According to this, if we can find $u^* \in \partial\varphi(u)$, then this element defines the subdifferential.

5) $u^* = |u|^{m-2} u \in \partial\varphi(u)$

We want to determine $u^* \in \varphi(u)$ for the phase space $H = H^{-1}(\Omega)$, i.e., by Definition 1.3 we have to find $u^* \in H_0^1(\Omega)$ such that

$$\langle u^*, v - u \rangle \leqslant \varphi(v) - \varphi(u) \qquad \forall v \in H^{-1}(\Omega). \tag{1.32}$$

Again we only need to study $u \in L^m(\Omega)$, and since (1.32) with $v \in H^{-1}(\Omega) \backslash L^m(\Omega)$ holds for every $u^* \in H_0^1(\Omega)$, it is also sufficient to take $v \in L^m(\Omega)$. Hence in fact we treat (1.32) in $L^m(\Omega)$, where we want to find $u^* \in L^{\frac{m}{m-1}}(\Omega) = L^m(\Omega)^*$ with

$$\int_\Omega u^*(v - u)\, dx \leqslant \frac{1}{m} \left(\|v\|_m^m - \|u\|_m^m \right) \qquad \forall v \in L^m(\Omega) . \tag{1.33}$$

In the following we will show that $u^* = |u|^{m-2}u \in L^{\frac{m}{m-1}}(\Omega)$ for $u \in L^m(\Omega)$ satisfies the desired property in $L^m(\Omega)$.

It holds

$$\int_\Omega |u|^{m-2}u(v - u)\, dx = \int_\Omega |u|^{m-2}uv\, dx - \int_\Omega |u|^m\, dx ,$$

which can be estimated by applying Hölder's and Young's inequality, both with $p = m$ and its conjugate $q = \frac{m}{m-1}$, as follows:

$$\int_\Omega |u|^{m-2}uv\, dx - \int_\Omega |u|^m\, dx$$

$$\leqslant \left\| |u|^{m-2}u \right\|_{\frac{m}{m-1}} \|v\|_m - \|u\|_m^m$$

$$\leqslant \frac{m-1}{m} \left\| u^{m-1} \right\|_{\frac{m}{m-1}}^{\frac{m}{m-1}} + \frac{1}{m} \|v\|_m^m - \|u\|_m^m$$

$$= \frac{m-1}{m} \|u\|_m^m + \frac{1}{m} \|v\|_m^m - \|u\|_m^m = \frac{1}{m} \left(\|v\|_m^m - \|u\|_m^m \right) .$$

Hence (1.33) is fulfilled and we have $|u|^{m-2}u \in \partial\varphi(u)$ in $L^m(\Omega)$.

In summary, for the subdifferential $\partial\varphi$ in $L^m(\Omega) \subset H^{-1}(\Omega)$ we were able to show

$$\partial\varphi(u) = |u|^{m-2}u \qquad \forall u \in L^m(\Omega) .$$

6) *Identifying $\tilde{u} = -\Delta_x(|u|^{m-2}u) \in \partial\varphi(u)$ for the phase space $H^{-1}(\Omega)$*

Thus our aim now is to determine the corresponding subdifferential in $H^{-1}(\Omega)$. For that we need to choose an appropriate scalar product on $H = H^{-1}(\Omega)$. This can be done by defining

$$(u, v)_H := \langle (-\Delta_x)^{-1}u, v \rangle \qquad \forall u, v \in H ,$$

where $\langle \cdot, \cdot \rangle$ denotes the duality pairing $_{H_0^1(\Omega)}\langle \cdot, \cdot \rangle_{H^{-1}(\Omega)}$. This definition is justified by the isomorphism $-\Delta_x : H_0^1(\Omega) \rightarrow H^{-1}(\Omega)$ provided by the famous Lax-Milgram Theorem. In accordance the inverse operator $(-\Delta_x)^{-1}$: $H^{-1}(\Omega) \to H_0^1(\Omega)$ exists and inherits the property of being self-adjoint.

Our purpose now is to find the identification $\tilde{u} \in H^{-1}(\Omega)$ of $u^* = |u|^{m-2}u$. For every element $\tilde{u}$ of $H^{-1}(\Omega)$ we have in accordance with the subdifferential definition

$$(\tilde{u}, v)_H = \langle (-\Delta_x)^{-1}\tilde{u}, v \rangle \stackrel{!}{=} \langle u^*, v \rangle \quad \forall v \in H \quad \Leftrightarrow \quad \tilde{u} = -\Delta_x u^* \in H^{-1}(\Omega).$$

$\square$

References

1. R.A. Adams, *Sobolev Spaces* (Academic, New York, 1975)
2. H. Brezis, *Functional Analysis, Sobolev Spaces and Partial Differential Equations* (Springer, New York, 2010)
3. G. Choquet, *Topology* (Academic, New York, 1966)
4. M. Efendiev, *Evolution Equations Arising in the Modelling of Life Sciences*, vol. 163 (Springer, Basel, 2013)
5. M.-A. Krasnosel'skii, A.-H. Armstrong, J. Burlak, *Topological Methods in the Theory of Nonlinear Integral Equations* (Pergamon, Oxford, 1964)
6. S.M. Nikol'skii, *Approximation of Functions of Several Variables and Imbedding Theorems*, vol. 205 (Springer, New York, 2012)
7. L. Nirenberg, *Topics in Nonlinear Functional Analysis*, vol. 6 (American Mathematical Society, Providence, 1974)
8. I.V. Skrypnik, *Nonlinear Elliptic Boundary Value Problems*, vol. 91 (BSB Teubner, Leipzig, 1986)
9. H. Triebel, *Interpolation Theory, Function Spaces, Differential Operators*. 2nd rev. a. enl. ed., English. 2nd rev. a. enl. ed. (Barth, Leipzig, 1995), p. 532
10. D. Werner, *Funktionalanalysis, 7., Korrigierte und Erweiterte Auflage* (Springer, Heidelberg, 2011)
11. E. Zeidler, *Functional Analysis and Its Applications, I. Fixed Point Theorem* (Springer, New York, 1990)

Chapter 2
Biological Background

The subject of this book is the mathematical modeling of a biological process, the swelling of mitochondria as it is graphically depicted in Fig. 2.1. Mitochondria are often termed the cell's powerhouse due to their main function as energy supplier for almost all eukaryotic cells. In this book we will become acquainted with another process that is highly regulated by mitochondria, namely cell death.

2.1 The Mitochondrion

The number of mitochondria in a cell varies widely related to the specific energy consumption, from one single organelle up to several thousands in muscle cells or neurons. For the experiments mostly liver mitochondria are used, which come up to 22% of the cell volume [1]. It is known that mitochondria within cells are not distributed randomly but are divided into three main regions. This feature will be interesting for the in vivo model to be described in Sect. 3.3. There, Fig. 5.1 shows the organization of an eukaryotic cell, restricted to the cell compartments which are of interest for our purpose.

Structure

Figure 2.2 depicts the mitochondrial structure, which will now be described in detail. The function of all occurring mitochondrial components will be explained, where the colors correspond to those from the diagram.

▸ One significant attribute of mitochondria is the enclosing double membrane, namely the inner (IM) and outer membrane (OM). A specific characteristic of the IM is the peculiar way of folded appearance, which compartmentalizes

© Springer Nature Switzerland AG 2018
M. Efendiev, *Mathematical Modeling of Mitochondrial Swelling*,
https://doi.org/10.1007/978-3-319-99100-9_2

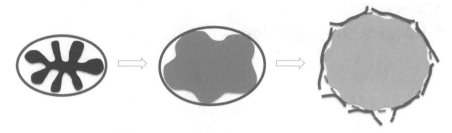

Fig. 2.1 The process of mitochondrial swelling: Extension of the inner membrane until it hits the outer membrane

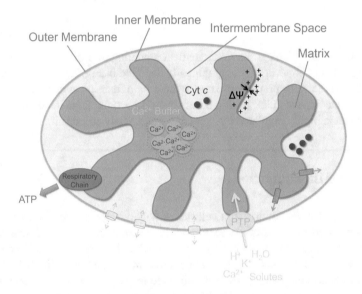

Fig. 2.2 Schematic description of the mitochondrial structure

it into numerous cristae. Since most of the important chemical reactions of mitochondria take place at its surface area, in doing so the potential working surface can be maximized. To that effect, mitochondria with higher energy demand exhibit more cristae and vice versa. In liver mitochondria, for instance, the surface of the IM exceeds that of the OM five times over [1].

As can be seen in the picture, the two membranes build up two compartments. These are the intermembrane space (IMS) and the matrix, each one with specialized functions to be presented below. The major difference between the mitochondrial membranes are different permeabilities.

The OM contains several protein channels, which allow for the exchange of molecules and ions up to a certain size [1]. According to this, the concentration of small molecules like ions and sugar in the IMS is nearly identical to that of the cytoplasm, whereas large molecules like proteins occur in much less amounts [13].

▹ By contrast, the IM is nearly impermeable to almost all molecules and so special membrane transporters are needed. These transporters include the calcium uniporter and several other ion exchange fluxes, see, e.g., [17]. As it is written in [12], the calcium uniporter plays a major role in intracellular Ca^{2+} signaling. Mitochondrial calcium uptake has controversial impacts on the mitochondrial as well as the cellular function. These are described in [6] and include control of the energy production rate or initiation of cell death. The latter, fatal property will be the topic of this book.

▹ Furthermore, mitochondria possess the ability to store a huge amount of calcium inside the matrix, the so-called Ca^{2+} buffer [14]. This calcium storage turns out to be of great importance for the mathematical modeling as it has some accelerating effect to mitochondrial swelling.

▹ ▹ For the maintenance of cellular respiration it is crucial to create an electro-chemical as well as concentration gradient at the IM by pumping protons from the matrix to the IMS [1]. Due to the impermeability of the IM, this leads to a proton gradient which is termed $\Delta_x \Psi$. The power of this gradient is utilized by the reflux of H^+ into the matrix through turbine-like channels. This flux produces energy which is then spent for the synthesis of ATP.

▹ During this process, electrons are transported by the small protein Cytochrome c (Cyt c). However, it also has an entirely different function. Under normal conditions, Cyt c cannot pass the OM. But if the OM is damaged or perforated by some reason, Cyt c is released from the IMS to the cytoplasm. This event is critical in cell death, since now apoptosis is inevitably triggered [8].

▹ As it was first mentioned in [9], there is also another way ions and solutes can enter the impermeable IM. Under pathological conditions, for example high Ca^{2+} concentrations, it happens that a special pore in the IM, the so-called permeability transition pore (PTP), opens. The PTP is formed connecting both membranes and has this name since pore opening makes the IM permeable. Later we will learn more about this pore and the serious consequences of its opening.

2.2 Apoptosis

Apoptosis is one of the most important types of programmed cell death. This phenomenon, first mentioned in a publication from 1972, can be described as a kind of "suicide program" of single cells, which have become ectopic or meaningless to the organism. Additionally, mutated or damaged cells use this mechanism to "sacrifice" themselves for the collective good and prevent further deteriorations [7].

The metamorphose of pollywogs to frogs and the degeneration of skins between fingers and toes during human embryonic development are famous examples of this sort of cell death.

The following information are taken from [7] and [13]. Apoptosis plays a crucial role in the maintenance of tissue homeostasis. At this, balance between an increasing cell population by proliferation and its decrease by cell death is required. The following data give an impression of the process dimension: without tissue homeostasis, an 80-year-old person would end up with two tons of bone marrow and lymph nodes together with a 16-km long gut. Apoptosis can be exogenously induced but it is enforced by the affected cell itself as a part of its metabolism.

In contrast to another fundamental cell death mechanism, namely necrosis, apoptosis underlies strong control procedures and assures the intactness of neighboring tissues. As opposed to it, necrosis effects cell swelling with subsequent destruction of the plasma membrane. This leads to local inflammations because released cytoplasm and organelles have to be removed by macrophages.

Pathways

Apoptosis can be elicited by several molecular pathways. The most essential ones, referred to as extrinsic and intrinsic pathways, are displayed in Fig. 2.3 and will be presented here. These information can be found in [4, 13].

Extrinsic pathway: On the right-hand side a schematical description of the extrinsic, also known as death receptor pathway, is given. At this juncture, the activation of death receptors at the cell surface assembles the death-inducing signaling complex (DISC). This complex in turn mobilizes several enzymes termed

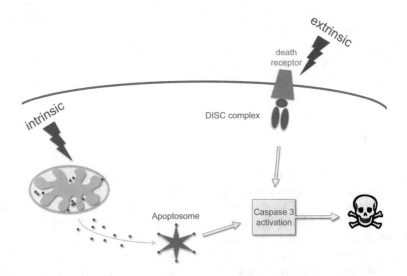

Fig. 2.3 Extrinsic and intrinsic pathways to caspase activation culminating in apoptosis

caspases which in the end cleave the effector caspase 3. Once these enzymes are formed, apoptosis is inexorably triggered.

Intrinsic pathway: In this book we focus on the intrinsic, also called mitochondrial pathway. Here apoptosis results from intracellular events, where mitochondria play an important role as can be seen on the left-hand side of Fig. 2.3. Mitochondrial stress is induced by several intracellular signals including high increase of the Ca^{2+} concentration within the cytoplasm, reactive oxygen species, DNA damage, toxins, or chemotherapy. It affects the mitochondrial membrane permeability and finally leads to the release of, among others, Cyt c, a common proapoptotic factor. A detailed description of the permeabilization process will be given in the next paragraph.

Cyt c binds to a special gene and thus elicits the formation of the so-called apoptosome complex. Due to its structure and effect, this complex is often termed the "wheel of death." The lethal function of the apoptosome is characterized by activation of caspase 3. At this stage the extrinsic and intrinsic pathways coincide and in both cases cell death by apoptosis is irreversibly initiated.

Mechanism: The activation of caspase 3, also called execution pathway, results in controlled cell destruction including cell shrinkage and DNA fragmentation. In the end the cell is fragmented into small apoptotic bodies, which in turn are digested by phagocytes. This ensures the clean and tidy removal of apoptotic cells from tissues and with that avoids the problems occurring at necrotic cell death, for instance inflammation.

2.3   Mitochondrial Permeability Transition (MPT)

In the past, mitochondria were only perceived as the cells powerhouse without any further role in the cell mechanism. Therefore it was very astonishing when it came to light that they also play a decisive role in the control of cell death. In the following we will point out the underlying mechanistic details of this important detection in conformity with [13].

As it was described before, the IM is usually nearly impermeable to all ions. However, this impermeability is not an enduring attribute. Under certain conditions the membrane can be permeabilized to solutes up to a certain size. This IM permeabilization is triggered by multiple factors, one of those being the topic of this book, the Ca^{2+}-induced mitochondrial permeability transition (MPT).

Permeability Transition Pore (PTP)

High loads of Ca^{2+} within the cytoplasm effectuate the permanent opening of the PTP introduced in the description of mitochondria. Consequently, due to the concentration gradient between IMS and matrix, positively charged ions like Ca^{2+}

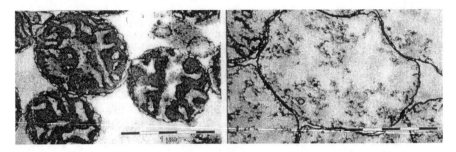

Fig. 2.4 Electron microscopy picture of unswollen (left) and swollen (right) mitochondria, picture taken at the Institute of Toxicology, Helmholtz Zentrum München

and H^+ are forced into the IMS. This equalization in charge and concentration immediately causes a collapse of the existing membrane potential $\Delta_x \Psi$. In turn this leads to an increase of the inner membrane permeability and with it to an osmotically driven influx of water and other solutes [5, 10].

As a natural result, the matrix starts to swell and the IM extends further and further until it hits the OM [3]. Due to the surface area of the IM largely exceeding that of the OM, the outer one gets even more permeable and in the end it ruptures. These occurrences can be visualized via an electron microscope as it is shown in Fig. 2.4. This OM permeabilization denotes a point of no return, since it enables the irreversible release of soluble proteins from the IMS. This is a critical event in cell death, because several proapoptotic factor including Cyt c are set free in this process. That means, once the OM of a sufficient amount of mitochondria is damaged, apoptosis is triggered and the cell will end up in death.

As it is, e.g., shown in [15], the rate of PTP activation is dependent on the actual Ca^{2+} load with higher calcium concentrations leading to faster pore opening. Also the experimental data (see Fig. 2.5) clearly reveal a positive dependence of the swelling mechanism to increasing concentrations.

Ca^{2+} Release

At the beginning we introduced the ability of mitochondria to store calcium inside the matrix. These Ca^{2+} buffers control the calcium homeostasis of the cell and contain large amounts of bounded calcium. If MPT is induced, the total matrix content of calcium is released containing the original buffer amount and the additionally assimilated Ca^{2+} [13].

For that reason, the remaining intact mitochondria are now confronted with an even higher load of Ca^{2+}. In fact, this leads to even faster pore opening and thus to an acceleration of the whole process. This mechanism is often termed "positive feedback" and, as it turns out, is of major importance for the mathematical modeling.

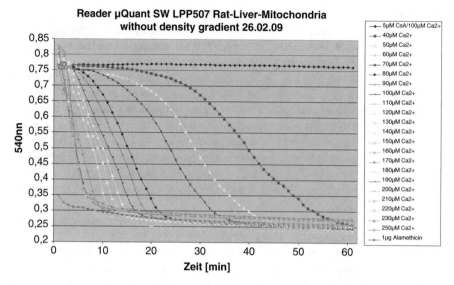

Fig. 2.5 Experimental data of mitochondrial swelling represented by the decrease of optical density

Pharmaceutical Background

Apoptosis is also of great interest for the pathophysiological research. It is recognized that this kind of programmed cell death contributes to many diseases in two oppositional ways, which are described in detail in [7, 13]. On the one hand, too much apoptosis is involved at (neuro-)degenerative diseases, Parkinson, Alzheimer, and AIDS, whereas on the other hand cancer and hemolytic anemia can be associated with too little of it.

Hence, it is not surprising that one of the most promising pharmacological strategies on cancer research is the therapeutic control of MPT. Triggering mitochondrial membrane permeabilization on cancer cells could be an excellent possibility to initiate apoptosis or at least overcome chemotherapeutic resistancies. In contrast to initiating apoptosis on virulent cells, pharmacological interventions can also be used to inhibit cell death as it is done, for instance, at ischemia/reperfusion injury. In this context, the main goal is to stabilize mitochondrial membranes to protect them from permeabilization.

Experimental Procedure

For reasons further elaborated above, MPT induces osmotical swelling of the mitochondrial matrix and with that causes an alteration of the molecular composition.

This modification can be quantified by the resulting change of the optical density. The research group of Hans Zischka from the Institute of Molecular Toxicology and Pharmacology at the Helmholtz Zentrum München performed all the experiments and provided the light scattering data [18].

Taking a look at Fig. 2.4, it becomes obvious that a population of intact mitochondria is a very heterogenous one. Light is deflected at the highly folded inner membranes to a great extent, which means we measure high light scattering values. Opposed to it, the more mitochondria are swollen, the more homogenous the population gets, which in turn leads to lower values. Like that mitochondrial swelling is measured indirectly by decreasing light scattering values. In agreement with findings reported in [2] and [15], we have determined this relation to be linear by use of free flow electrophoresis, a technique to partition mitochondria that have undergone MPT [18].

The effects of MPT are measured by use of an absorbance reader that yields optical density data according to the curves displayed in Fig. 2.5.

Here the swelling curves depict the mean value of four independent light scattering measurements. For that, a microplate with 4×24 repositories is filled with identical mitochondrial amounts each. Afterwards these are treated with 24 substances, for example different Ca^{2+} concentrations, in order to obtain four independent measurements of similar mixtures. The absorbance reader then quantifies the corresponding light scattering data at a wavelength of 540 nm. These experimental data show that the more Ca^{2+} is added, the faster the whole swelling proceeds. Starting from time ranges of about 1 h, at very high concentrations swelling is completed after less than 10 min. The swelling curve appearance is similar for each Ca^{2+} concentration, with an initial lag phase followed by a steep decrease of optical density. This initial phase of moderate decay can be explained by the duration of calcium uptake and the time it needs until the permeabilization process is initiated.

However, there is a 1 min time lag between the substance addition on all repositories and the start of the measurement. Indeed, one has to keep in mind that at high Ca^{2+} concentrations the swelling proceeds very fast and thus one missing minute implies a high loss of information. This fact poses problems for the mathematical modeling and with that, the faster swelling proceeds the worse the approximation of mitochondrial incidents gets.

The required mitochondria are extracted from rat liver, isolated from debris and nuclei by multiple centrifugations and resuspended in an isolation buffer. The intactness of the organelles prepared like this is tested by measuring the respiration activity in form of oxygen consumption. On intact mitochondria, osmotic swelling is induced using the "standard swelling buffer" consisting of 10 mM MOPS-Tris, pH 7.4, 200 mM sucrose, 5 mM succinate, 1 mM Pi, 10 μM EGTA, and 2 μM rotenone [18].

In the experiments, the common MPT inducer Ca^{2+} is used. As reported in [11] and [16], there are many indicators that calcium plays a main role in several forms of apoptosis, even when MPT is initiated by other substances. This fact traces back to the huge amount of stored calcium in the endoplasmic reticulum, which is then released and targets the mitochondrial membrane permeability. Hence it is really important to obtain a deep understanding of the calcium induced swelling mechanism.

References

1. B. Alberts, D. Bray, K. Hopkin, A. Johnson, J. Lewis, M. Raff, K. Roberts, P. Walter, *Essential Cell Biology* (Garland Science, New York, 2013)
2. S.V. Baranov, I.G. Stavrovskaya, A.M. Brown, A.M. Tyryshkin, B.S. Kristal, Kinetic model for Ca2+-induced permeability transition in energized liver mitochondria discriminates between inhibitor mechanisms. J. Biol. Chem. **283**(20), 665–676 (2008)
3. G.A. Blondin, D.E. Green, The mechanism of mitochondrial swelling. Proc. Natl. Acad. Sci. **58**(2), 612–619 (1967)
4. C. Bortner, J. Cidlowski, Apoptotic volume decrease and the incredible shrinking cell. Cell Death Differ. **9**(12), 1307–1310 (2002)
5. G. Calamita, D. Ferri, P. Gena, G.E. Liquori, A. Cavalier, D. Thomas, M. Svelto, The inner mitochondrial membrane has aquaporin-8 water channels and is highly permeable to water. J. Biol. Chem. **280**(17), 17149–17153 (2005)
6. M.R. Duchen, Mitochondria and calcium: from cell signalling to cell death. J. Physiol. **529**(1), 57–68 (2000)
7. S. Elmore, Apoptosis: a review of programmed cell death. Toxicol. Pathol. **35**(4), 495–516 (2007)
8. D.S. Goodsell, The molecular perspective: cytochrome C and apoptosis. Oncol. **9**(2), 226–227 (2004)
9. D.R. Hunter, R.A. Haworth, The Ca2+-induced membrane transition in mitochondria: I. The protective mechanisms. Arch. Biochem. Biophys. **195**(2), 453–459 (1979)
10. D.R. Hunter, R. Haworth, J. Southard, Relationship between configuration, function, and permeability in calcium-treated mitochondria. J. Biol. Chem. **251**(16), 5069–5077 (1976)
11. G. Kass, S. Orrenius, Calcium signaling and cytotoxicity. Environ. Health Perspect. **107** Suppl 1, 25 (1999)
12. Y. Kirichok, G. Krapivinsky, D.E. Clapham, The mitochondrial calcium uniporter is a highly selective ion channel. Nature **427**(6972), 360–364 (2004)
13. G. Kroemer, L. Galluzzi, C. Brenner, Mitochondrial membrane permeabilization in cell death. Physiol. Rev. **87**(1), 99–163 (2007)
14. D.G. Nicholls, S. Chalmers, The integration of mitochondrial calcium transport and storage. J. Bioenerg. Biomembr. **36**(4), 277–281 (2004)
15. V. Petronilli, C. Cola, S. Massari, R. Colonna, P. Bernardi, Physiological effectors modify voltage sensing by the cyclosporin A-sensitive permeability transition pore of mitochondria. J. Biol. Chem. **268**(29), 21939–21945 (1993)
16. P. Pinton, C. Giorgi, R. Siviero, E. Zecchini, R. Rizzuto, Calcium and apoptosis: ER-mitochondria Ca2+ transfer in the control of apoptosis. Oncogene **27**(50), 6407–6418 (2008)
17. A.V. Pokhilko, F.I. Ataullakhanov, E.L. Holmuhamedov, Mathematical model of mitochondrial ionic homeostasis: three modes of Ca 2+ transport. J. Theor. Biol. **243**(1), 152–169 (2006)
18. H. Zischka, N. Larochette, F. Hoffmann, D. Hamoller, N. Jagemann, J. Lichtmannegger, L. Jennen, J. Muller-Hocker, F. Roggel, M. Gottlicher et al., Electrophoretic analysis of the mitochondrial outer membrane rupture induced by permeability transition. Anal. Chem. **80**(13), 5051–5058 (2008)

Chapter 3
Model Description

The process of mitochondrial swelling induced by MPT is known for more than 35 years. However, many important issues concerning the MPT have still remained unanswered or controversial. It is for instance a matter of continuous debate which components exactly build up the PTP [3, 7]. Moreover, up to now we do not have data of the process in a living cell and thus the mathematical models considered in this book can help to understand the swelling of mitochondria for the biologically and especially pharmacologically more relevant case. For that reason mathematical modeling is of great importance. Indeed, it provides the possibility to verify and predict properties of the underlying biological mechanism that possibly cannot be obtained from the experiments directly.

3.1 Existing Models

The overview in this section is mainly adapted from [6]. In order to improve the understanding of the kinetics and the complex interdependences of the MPT process, modeling of the MPT pore function has only started recently with two conceptually different approaches.

Microscale

One is mainly oriented on a detailed biochemical and biophysical description of mitochondrial molecular processes such as mitochondrial respiration or ion exchanges [12, 13]. For each of these processes an equation is created, which are then combined in a system of nonlinear ordinary differential equations including a

© Springer Nature Switzerland AG 2018
M. Efendiev, *Mathematical Modeling of Mitochondrial Swelling*,
https://doi.org/10.1007/978-3-319-99100-9_3

number of variables, e.g., the amount of Ca^{2+} inside the matrix, the pH-value, or the membrane potential $\Delta_x \Psi$.

The specific advantage of this approach is that it can reproduce the three states of the pore: closed, flickering, or permanently open. However, this model does not display the time course of pore opening and lacks a major feature, the irreversible volume increase. Hence it is inadequate for simulating mitochondrial swelling. Furthermore, this kind of model only considers the processes for single mitochondria, whereas the experiments are made with huge mitochondrial populations where the resulting data always represent a mean value.

Macroscale

The other modeling approach aims to directly represent mitochondrial swelling. In contrast to the microscale models, it considers a population of mitochondria and studies the total volume increase. It focuses on the basic kinetic processes and hence is mathematically and numerically comparatively easier to handle. It consists of only one [10] or two [1] equations and concentrates on the increase of the number of swollen mitochondria, largely ignoring the details of the underlying biochemical mechanism. Despite these simplifications, this approach can produce a more accurate picture of the mitochondrial volume increase, which can be directly compared with the experimental data.

First Order Kinetics

To our knowledge, Massari [10] created the first model of this kind assuming first order kinetics. A great advantage of this model is that it can be solved explicitly due to its low mathematical complexity. A drawback of this model is that it fails to account for the initial lag phase in mitochondrial swelling displayed in Fig. 2.5. In agreement with the observations mentioned in [1] and [10], we have observed that the Massari model especially fits the end, the "tail" of the swelling curves, but misses their starting phase. The reason for this is the assumption in the derivation of the model, according to which the logarithm of the mitochondrial volume changes linearly during the swelling process. In fact, we showed that for our experimental data this linearity only occurs once most of the actual swelling is done.

Several Steps of Calcium Uptake

Baranov et al. [1] then presented an elaborate model, which provides a good simulation of the swelling on a longer time interval. It consists of two ODEs, one for the amount of calcium and one for the ratio of swollen mitochondria. The authors take into consideration that the Ca^{2+} uptake by mitochondria occurs in several steps with different reaction rates. However, their simulations concentrate on the middle part of the experimental swelling curves, and unfortunately do not explain the above-mentioned "tail." The change of the parameters is tested and discussed in dependence of various inducers and inhibitors, but the variation of parameter values with increasing amount of added Ca^{2+} was not examined.

Second Order Kinetics

Due to the above-mentioned disadvantages of the existing models, our aim was to develop a mathematical model that is capable of simulating the whole swelling process. The derived model is based on the known major properties of the process, which allows us to analyze the parameter values and obtain a classification of different swelling inducers or mitochondria from different tissues. In the following we will briefly present this model, which can be found in [6].

Based on experiments we assume that three subpopulations of mitochondria with different corresponding volumes exist: unswollen, swelling, and mitochondria that have completely finished swelling. The first and the last group have constant mean volumes, depending only on their source and the medium. The mean volume of swelling mitochondria additionally depends on the characteristics of the swelling process, which could be influenced, e.g., by properties and concentrations of the added substances. Onset and time course of swelling typically vary between mitochondria from different tissues, caused, e.g., by different sensitivities to inducers [2, 4].

The dynamical behavior of the total volume of the mitochondrial population, i.e., the subsumed volume of all subpopulations corresponds to the light scattering data obtained from the experiments. As noted before, we have determined the relation of volume increase and optical density decrease to be linear.

Model Description The model is based on the observation that mitochondria vary concerning their sensitivity for swelling induction by stimuli like Ca^{2+} as it was described by our collaboration partner in [14]. Below we follow [6], where the authors modeled the time progress of swelling with two equations for two variables X and V.

Here $X(t)$ denotes the fraction of mitochondria that are swollen or have started swelling at time t and hence $0 \leqslant X(t) \leqslant 1$. $V(t)$ describes the average volume of the mitochondrial population at time t. Since the authors assume that all mitochondria are intact prior to calcium addition, it holds $X(0) = 0$. Let X_p be the ratio of swollen mitochondria after the whole swelling is done. During the extraction and isolation from living mitochondria, it happens that some mitochondria are destructed and hence cannot react to swelling inducers. This is why it holds $0 < X_p \leqslant 1$ and in particular we assume $X_p = 0.9$ corresponding to 10 % loss in accordance with experimental observations. By V_0 and V_p we denote the volume of unswollen and completely swollen mitochondria, respectively. These parameter values are chosen in agreement with [12] and we take $V_0 = 1.2$ and $V_p = 1.7$ with unit [ml/mg protein].

The permeability transition process can be described via the initial value problem

$$\frac{d}{dt}X(t) = (aX(t) + b)(X_p - X(t))$$

$$X(0) = 0,$$

where $a \geqslant$ and $b > 0$ are biological parameters.

Here a describes the positive feedback induced by stored calcium additionally being released when mitochondria get completely swollen. The value $a = 0$ corresponds to the case of no feedback and we are in the situation of the linear kinetics model described earlier. However, it turns out that for an appropriate simulation of the whole swelling process a is always positive, and thus we confirmed the existence of such an accelerating effect mathematically. Parameter b is a background swelling coefficient, i.e., it represents the swelling rate which is induced by the starting stimulus, in our case the addition of Ca^{2+}.

This second order ODE can be solved explicitly by separation of variables, which yields

$$X(t) = X_p \left(1 - \frac{X_p + \frac{b}{a}}{X_p + \frac{b}{a}\exp\left((aX_p + b)t\right)} \right).$$

This representation shows that $X(t)$ is monotone increasing with one inflection point and that

$$\lim_{t \to \infty} X(t) = X_p.$$

The solution is also robust in the sense that the rate of convergence is exponential.

In [6] the authors split up the mitochondrial volume $V(t)$ into three different subpopulations V_1, V_2, and V_3. With the delay term $\tau > 0$ denoting the average swelling time of a single mitochondrion, we have

$V_1(t) = (1 - X(t))V_0$

$\qquad$ (3.1)

$\qquad$ Mitochondria that did not yet start swelling at time t

$V_2(t) = (X(t) - X(t - \tau))kV_p, \quad 0 < k < 1$

$\qquad$ Mitochondria that have started swelling after $t - \tau$ and hence are not

$\qquad$ fully swollen at time t; they only come up to a fixed percentage k

$\qquad$ of the final volume

$\qquad$ (3.2)

$$V_3(t) = X(t - \tau)V_p$$

(3.3)

Mitochondria that finished swelling completely until $t = t - \tau$

By setting

$$X(t) := 0 \quad \text{for } t < 0,$$

we obtain the volume equation

$$V(t) = (1 - X(t))V_0 + (X(t) - X(t - \tau))kV_p + X(t - \tau)V_p.$$

(3.4)

The parameter k determines the average swelling volume of mitochondria that are in the swelling process. Naturally, k cannot be arbitrarily close to 0, it has a lower bound depending on the experimental setting.

Assuming the average volume to be kV_p necessarily leads to a small break of the curve at $t = \tau$. This results from the fact that the right-hand side derivative of $X(t)$ at $t = 0$ is $\lim_{t \to 0+} X'(t) = bX_p$ and hence $\lim_{t \to \tau+} V'(t) - \lim_{t \to \tau-} V'(t) = bX_p V_p(1 - k) > 0$. Nevertheless, the size of the break tends to 0 as $b \to 0$. As most of our experiments showed rather low and almost invariant values for b, we could largely eliminate the break.

Results

1. *Parameter estimation:*

Figure 3.1 shows the experimental data and the rescaled simulated volume curves for various Ca^{2+} concentrations. Here the four parameters a, b, τ, and k are estimated by means of least squares minimization using the Nelder-Mead simplex method [9]. Like that we found an accurate fit of the volume curves and the measurement for the total time range of swelling. Moreover, the model parameters change in a well-determined way, consistent with the corresponding Ca^{2+} concentrations.

As already mentioned, the background swelling parameter b shows no clear correlation with the amount of added calcium over a wide range and always remains around a value of 0.021. Therefore it is possible to fix this parameter in order to reduce the complexity of minimization. The feedback parameter a is estimated to be always positive, i.e., indeed leading to an accelerating effect. Furthermore, by assuming second order kinetics, it was possible to appropriately describe the initial lag phase as well as the "tail" of the swelling curve. As a result a linear increase of a with increasing Ca^{2+} concentrations was determined.

The average swelling time τ decreases exponentially with increasing amount of added calcium. This behavior is probably governed by signal transfer processes: a high amount of extra-mitochondrial Ca^{2+} results in a faster arrival of calcium ions inside the mitochondria and hence in a faster initiation of MPT.

Parameter k represents the mean value of mitochondria during their swelling process. Higher values of k thus indicate a faster volume increase in the

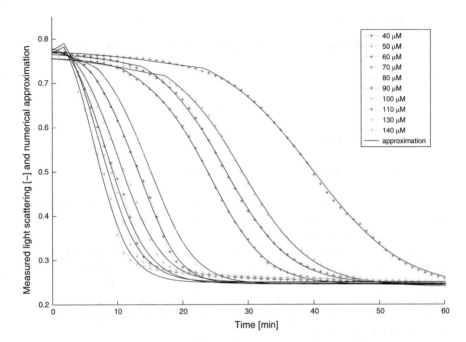

Fig. 3.1 Swelling curves at different Ca^{2+} concentrations compared to the corresponding numerical approximation assuming second order kinetics (with permission from BMC Research)

beginning of the swelling compared to the end, e.g., a less convex/more concave swelling curve of single mitochondria. For our data, k was estimated to be around 0.75 with a very slow exponential increase dependent on the added calcium concentration.

Example 3.1

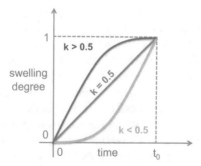

In contrast to the population volume between V_0 and V_p, for simplicity we take a look at single mitochondria with a swelling degree between 0 and 1. Then $k = 0.5$ corresponds to linear swelling curves. Higher values of k lead to faster than linear swelling, whereas lower values represent a slower volume increase. The influence of single swelling curves to the total mitochondrial volume was studied in [5].

2. *Curve progression:*

The shape of the resulting volume curves highly depends on the choice of parameter k as it represents the swelling progression of single mitochondria. As was shown for example in [2] or [11], at high Ca^{2+} concentrations mitochondria may go through a short phase of initial shrinking before the actual swelling begins, i.e., we have a slow increase of optical density in the very beginning. Obviously, a value of k with $kV_p < V_0$ means that mitochondria first shrink, thus loose from the initial volume V_0, and then a fast swelling follows to reach the final volume V_p. However, $kV_p > V_0$ does not exclude initial shrinking, but indicates a domination of partly swollen mitochondria over shrunk ones.

Furthermore, for very high values of k we observe a different curve progression, which has a two-phase behavior consisting of two inflection points.

3. *Other organs, different inducers:*

Figure 3.2 presents the comparison of the swelling curves obtained by treating liver and kidney mitochondria with the same amount of Ca^{2+}.

It is obvious that the swelling curves have completely different curve progressions, but nevertheless this model produces very accurate results. As mentioned before, the two-phase behavior of the kidney swelling curve can be simulated by high values of parameter k. The optimal parameter values are given as follows:

Liver mitochondria $50\mu M\ Ca^{2+}$: $\tau = 16.28$ $a = 0.25$ $b = 0.012$ $k = 0.74$
Kidney mitochondria $50\mu M\ Ca^{2+}$: $\tau = 24.65$ $a = 0.25$ $b = 0.019$ $k = 0.89$.

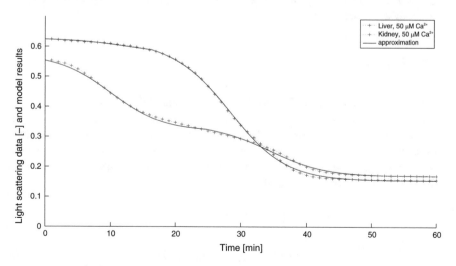

Fig. 3.2 Different curve progressions of liver and kidney mitochondria exposed to the same amount of Ca^{2+} (with permission from BMC Research)

Hence the difference results from differing values of τ and k, while the optimal values for a and b are almost identical. This refers to varying Ca^{2+} uptake and corresponding swelling times, but similar feedback mechanisms in the two organs.

In [6] a comparison of liver mitochondria with swelling induced by $50\,\mu M$ Ca^{2+} and $20\,\mu M$ Hg^{2+} was made. This gave the following optimal parameters:

Liver mitochondria $20\mu M\ Hg^{2+}$: $\tau = 16.54$ $a = 0.055$ $b = 0.041$ $k = 0.70$.

Here all parameters in comparison with liver mitochondria exposed to Ca^{2+} are almost the same, except for the feedback rate a. This definitely makes sense, since this model takes a look at identical mitochondria, which by default should have the same swelling time and speed. Note that the mechanism of positive feedback is connected to calcium and since in [6] the authors induce swelling by mercury, this effect cannot appear to the same extent anymore. However, it is not completely zero, because if a sufficiently high amount of mitochondria is swollen, then the stored Ca^{2+} itself can induce swelling, even if it was induced by Hg^{2+}.

In summary, the model of mitochondrial swelling developed in this section is capable to describe the experimental data accurately and over the whole time range. Furthermore it is not only used to reproduce given data, but we are now also able to classify mitochondria and inducers by the corresponding parameter values. If we think for example of the swelling curves for liver and kidney mitochondria, then from the data itself it is not possible to deduce what causes these different curve shapes. But by means of parameter τ we can state that kidney mitochondria have a longer average swelling time.

3.2 Spatial Effects

Coming from an ODE model, it is natural to think about the necessity of including spatial effects by means of taking into account partial differential equations. In Sect. 3.1, we introduced a model which is only focused on the time evolution. That means we do not take into account local effects and only work with mean values over the whole domain. In other words, the three volume components described in (3.1)–(3.3) can be written as weighted integrals:

$$V_1(t) = V_0 \int_{\Omega} N_1(x,t)\,dx$$

$$V_2(t) = kV_p \int_{\Omega} N_2(x,t)\,dx$$

$$V_3(t) = V_p \int_{\Omega} N_3(x,t)\,dx$$

where N_1, N_2, and N_3 denote the density of unswollen, swelling, and completely swollen mitochondria. The evolution of these densities depends on the local Ca^{2+} concentration, which is denoted by $u(x, t)$. The added amount of calcium as inducer of mitochondrial swelling is hence given by the initial data $u_0(x)$.

Remark 3.1 Here the question appears how a density $N(x)$ of mitochondria on the domain Ω is defined. Mitochondria are a discrete quantity, which has to be translated into a continuous density. Here we present the easiest way to obtain such a density function:

First we define a lattice Γ of points covering Ω. Then for every point $x_i \in \Gamma$ and fixed radius $\varepsilon > 0$ we draw a circle $B_\varepsilon(x_i)$ and count the number n_i of mitochondria within this circle. We set $N(x_i) = n_i$, then interpolation yields a continuous function $N(x)$ for every $x \in \Omega$. In order to derive a density, we have to divide N by the size of Ω.

We assume we are given a test tube with purified mitochondria. If there were no spatial effects, then it should make no difference how the same amount of calcium is added. That is, as long as the integral $\int_\Omega u_0(x)\, dx$ is equal, the resulting mitochondrial volume

$$V(t) = V_1(t) + V_2(t) + V_3(t)$$

should be the same. Figure 3.3 depicts different possibilities to add the same Ca^{2+} amount. The following experiment reveals that different distributions indeed have an influence to the volume outcome. Here three settings with the same total calcium amount are considered:

1. **Volume ratio 1:4**

   The volume ratio of added Ca^{2+} compared to the volume of the total (unswollen) mitochondrial population is 1:4, i.e., the calcium source is near to being uniformly distributed. This setting is the standard setting for the experimental data displayed in Fig. 2.5.

2. **Volume ratio 1:100 mixed**

   Here the calcium appears in a much more concentrated form, in order to obtain the same calcium amount in a much smaller volume. After addition to

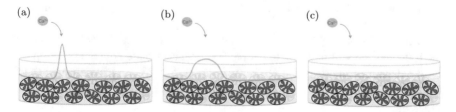

Fig. 3.3 Addition of a fixed calcium amount in varying distribution. (**a**) Highly localized, (**b**) "normally distributed," (**c**) uniformly distributed

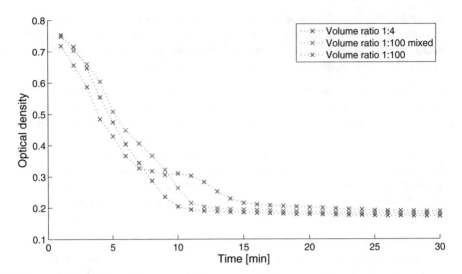

Fig. 3.4 Experimental data resulting from different volume ratios of Ca^{2+} and mitochondria

the mitochondria the whole content is mixed, which again leads to a fast calcium dispersion.

3. **Volume ratio 1:100**

Calcium is added in a high concentration with ratio 1:100, but this time no mixing takes place. That means we are in the highly localized case.

Figure 3.4 shows the results of these experiments. Here it becomes obvious that the swelling curves and with that the volume increase function $V(t)$ highly depend on the initial calcium distribution. Mixing the test tube content of mitochondria and added calcium leads to a faster dispersion of the initially highly localized Ca^{2+} source and hence we soon arrive at the nearly uniformly distributed case. This explains why the blue and green curves show almost the same progression.

However, if we do not mix and wait until calcium is distributed by diffusion, the resulting red curve shows a completely different shape. In the beginning the swelling proceeds faster due to locally very high calcium concentrations, but at the same time several mitochondria remain untreated. Hence, in contrast to the other cases, it takes some time until all of them are reached, which explains the slower second phase of swelling.

With that the necessity of taking spatial effects into account is motivated and in the following we will focus on this topic. That means we are now interested in the local behavior of the densities $N_1(x, t)$, $N_2(x, t)$, and $N_3(x, t)$ in dependence on the calcium concentration $u(x, t)$ instead of only taking a look at the mean values of them.

In order to develop a mathematical model, we have to consider two spatial effects that directly influence the process of mitochondrial swelling:

On the one hand, the extent of mitochondrial damage due to calcium is highly dependent on the position of the particular mitochondrion. If the mitochondrion is located near to the Ca^{2+} source, it is exposed to a higher dose compared to mitochondria residing further away and consequently it will be damaged to a higher degree. By diffusion, the locally high calcium dose is diminished and the remaining mitochondria are confronted with a lower concentration.

On the other hand, at a high amount of swollen mitochondria the effect of the positive feedback gets relevant and here the residual mitochondria are confronted with a higher calcium load. Due to the positive feedback mechanism and the natural diffusion, we have some kind of spreading calcium wave, e.g., reported in [8], which however is neglected in the existing models. Experiments of our collaboration partner confirmed that not all mitochondria are damaged to the same degree which implies that they do not react homogeneously [14].

This explains why the distribution of the added Ca^{2+} amount indeed leads to different swelling dynamics. Hence it is important to include the calcium evolution into the model and introduce spatial effects by means of calcium diffusion. This leads to a partial differential equation model.

Remark 3.2 The mathematical model to be developed on the following chapters does not aim to give a quantitative description of the experimental data as we obtained from the ODE model described earlier. The given data only have mean value character and hence from them we cannot deduce more than mean value information. However, our aim now is to qualitatively understand the underlying biological mechanism on the basis of local effects and in particular make the transfer to the processes taking place in vivo.

In Vitro Swelling

The experimental procedure introduced in Chap. 2 describes the set up for in vitro swelling. Here living mitochondria are extracted from organs and then swelling is artificially triggered within a test tube by addition of MPT inducing substances like Ca^{2+}. Naturally calcium cannot leave the test tube and we look at a process taking place in a closed system without any disturbance from the outside. The test tube contains a huge amount of purified mitochondria, which rest in a kind of sugar solution that prevents mitochondria from dying immediately. In this solution the mitochondria are uniformly and "densely" distributed. Furthermore they are either intact or have been destructed while extraction, but in any case at the beginning there are no swelling or completely swollen mitochondria in the test tube.

The initial distribution of calcium describes how it is added to the mitochondria. The experimental data are obtained by assessing a calcium to mitochondria ratio of 1:4, i.e., 20% of the test tube content then is Ca^{2+}. If one imagines the dissolving of ink in a glass of water, adding ink in a ratio of 1:4 leads to an almost immediate uniform distribution.

Therefore, in this case the effects of spatial dependencies are small, since all mitochondria are exposed to the almost same calcium amount simultaneously. Hence the positive feedback only occurs when a large part of the mitochondria is already completely swollen and with that the accelerating effect only happens once most of the swelling is finished.

$$u(x, t) : Ca^{2+}\text{concentration}$$
$$N_1(x, t) : \text{density of unswollen mitochondria}$$
$$N_2(x, t) : \text{density of mitochondria in the swelling process}$$
$$N_3(x, t) : \text{density of completely swollen mitochondria}$$

As we will see in the numerical simulations, the experimental data are best reproduced by assuming highly dissolved initial calcium concentrations, whereas more located initial data lead to completely different curve shapes. Hence, one really has to be aware of the influence of the experimental design on the swelling curve outcome. In order to obtain comparable results, one has to pay attention to what degree of localization the swelling inducer is added.

3.3   The Mitochondria Model: In Vitro

In the following the biological description of the mitochondrial swelling process is translated into the mathematical language. We already indicated before that for the in vitro model we choose the domain Ω to be the test tube, whereas the in vivo model considers Ω as the whole cell.

The Variables

The model variables were already introduced before and are described in the following way: Here the transition of intact mitochondria over swelling to completely swollen ones proceeds in dependence on the local calcium concentration. At this we can assume that mitochondria in the test tube as well as within cells do not move in any direction and hence the spatial effects are only introduced by the calcium evolution.

Initial Conditions

In accordance with previous findings, the initial data

$$u(x,0) = u_0(x)\,, \quad N_1(x,0) = N_{1,0}(x)\,, \quad N_2(x,0) = N_{2,0}(x)\,, \quad N_3(x,0) = N_{3,0}(x)$$

are defined by the type of experiment.

Here the quantities N_i, $i = 1,2,3$ are to be interpreted in the density sense described earlier when we scale $\frac{1}{|\Omega|} N_i(x,t)$. We assume that they initially take only the values 0 or 1, where $N_{i,0}(x) = 0$ corresponds to "no mitochondrion at point x" and a value of 1 means "mitochondrion at point x."

in vitro: Initially the test tube contains densely distributed mitochondria and the "swelling buffer" described in Chap. 2. It does not contain any calcium, hence we have

$$u_0(x) = u_{\text{peak}}(x) \geqslant 0\,, \quad N_{1,0}(x) \equiv 1\,, \quad N_{2,0}(x) \equiv 0\,, \quad N_{3,0}(x) \equiv 0\,.$$
$$(3.5)$$

Here u_{peak} describes the artificially added Ca^{2+} that induces the swelling process.

Note that in the beginning there are no swelling mitochondria or those that have already finished swelling.

Remark 3.3 For the mathematical analysis we allow for more general initial conditions.

Boundary Conditions

The spatial effects only trace back to spreading Ca^{2+}, hence we only need to impose boundary conditions for the variable $u(x,t)$:

in vitro: Calcium cannot cross the boundary $\partial\Omega$, i.e., the test tube wall and hence we have homogenous Neumann boundary conditions

$$\partial_\nu u(x,t) = 0 \quad \text{for } x \in \partial\Omega\,.$$

Mathematically we will also treat the case of homogenous Dirichlet boundary conditions

$$u(x,t) = 0 \quad \text{for } x \in \partial\Omega\,.$$

Biologically, this kind of boundary condition appears if we put some chemical material on the wall that binds calcium ions and hence removes it as a swelling inducer.

Now we are finally prepared to introduce the new model. As we noted before, only the calcium concentration u evolves in time and space, and hence its behavior is described by a partial differential equation. Mitochondria do not move and thus the evolution of the mitochondrial subpopulations N_1, N_2, and N_3 is modeled by a system of ordinary differential equations that are dependent on the current state of $u(x, t)$. These ODEs depend on the space variable x in terms of a parameter. This leads to the following coupled ODE-PDE system determined by the model functions f and g and the diffusion operator Δ_x, i.e., the standard Laplacian,

$$\partial_t u = d_1 \Delta_x u + d_2 g(u) N_2$$
$$\partial_t N_1 = -f(u) N_1$$
$$\partial_t N_2 = f(u) N_1 - g(u) N_2$$
$$\partial_t N_3 = g(u) N_2$$

with diffusion constant $d_1 \geqslant 0$ and feedback parameter $d_2 \geqslant 0$. The boundary conditions will vary depending on the type of the model and can be generally formulated as

$$a\,u + b\,\partial_\nu u = h \text{ on } \partial\Omega$$

with functions a, b, and h defined on the boundary. Furthermore we have initial conditions

$$u(x, 0) = u_0(x), \quad N_1(x, 0) = N_{1,0}(x), \quad N_2(x, 0) = N_{2,0}(x), \quad N_3(x, 0) = N_{3,0}(x)$$

with the specific functions depending on the model type as described earlier.

This model is solved on the phase space H, which has to be suitably defined for every type of problem. According to that, the initial data have to be taken from an appropriate space dependent on the diffusion operator A. For the ODE components we can always assume

$$N_{i,0} \in L^\infty(\Omega) \quad \text{for } i = 1, 2, 3.$$

In the following we will explain all model components in detail.

Model Function f

The process of mitochondrial permeability transition is dependent on the calcium concentration. If the local concentration of Ca^{2+} is sufficiently high, the transition pores open and mitochondrial swelling is initiated. This incident is mathematically described by the transition of mitochondria from N_1 to N_2. The corresponding

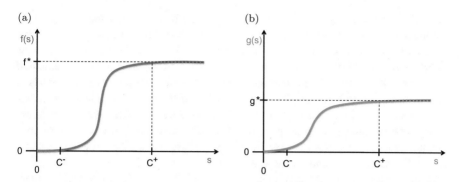

Fig. 3.5 Graphical description of the model functions we used for the numerical simulation. (**a**) Model function f. (**b**) Model function g

transition function $f(u)$ is zero up to a certain threshold C^- displaying the concentration which is needed to start the whole process. Whenever this Ca^{2+} threshold is reached, the local transition at this point from N_1 to N_3 over N_2 is inevitably triggered. It is written, e.g., in [11] that this process is calcium-dependent with higher concentrations leading to faster pore opening. Hence the function $f(u)$ is increasing in u.

The transfer from unswollen to swelling mitochondria is related to pore opening, hence we also postulate that there is some saturation rate f^* displaying the maximal transition rate. This is biologically explained by a bounded speed of pore opening with increasing calcium concentrations. This saturation with respect to calcium also becomes obvious in the experimental data displayed in Fig. 2.5, where we see that with increasing Ca^{2+} concentrations the swelling process is speeded up, but at very high concentrations we do not observe changes in the swelling curves appearance any more.

Figure 3.5a displays the type of function we used for the numerical simulations.

Remark 3.4 The initiation threshold of f is crucial for the whole swelling procedure. Dependent on the amount and location of added calcium, it can happen that in the beginning the local concentration was enough to induce swelling in this region, but after some time due to diffusion the threshold C^- is not reached anymore. Thus we only have partial swelling and after the whole process there are still intact mitochondria left. Nevertheless, there are no mitochondria in the intermediate state N_2.

Model Function g

The change of the population N_2 consists of mitochondria entering the swelling process (coming from N_1) and mitochondria getting completely swollen (leaving to N_3). The transition of N_2 to N_3 is modeled by the transition function $g(u)$. In contrast to the function f here we have no initiation threshold and this transition cannot be avoided (as we will see later: $u(x, t) > 0 \; \forall x \; \forall t$ except for the homogenous Dirichlet case). This property is based on the biological mechanism of swelling. The permeabilization of the inner membrane due to pore opening leads to an osmotically driven influx of water and other solutes into the mitochondrial matrix and induces swelling. This process itself is independent of the present calcium concentration. Due to a limited pore size, this effect also has its restrictions and thus we have saturation at level g^*. By simplicity we assume that the point of saturation C^+ is the same for f and g as can be seen in Fig. 3.5b, where we depict a typical function g used in the simulations. However, biologically it is not clear if there are other influences of calcium to this second transition, e.g., by the opening of additional pores. To include such possibilities, for the simulations we assume g to be increasing with saturation at level g^* for $u \geqslant C^+$.

The third population N_3 of completely swollen mitochondria grows continuously due to the unstoppable transition from N_2 to N_3. All mitochondria that started to swell will be completely swollen in the end.

Remark 3.5

1) For the mathematical analysis we allow for more general transition functions f and g. The corresponding assumptions are formulated in Conditions 4.1 and 4.2. Note that in particular f and g (up to a small interval) not even have to be monotone.
2) In the present model we assume that mitochondria only differ with respect to their location and with that in the moment when they are hit by the calcium wave.

   If we take a look at the whole cell, then it is not clear if all mitochondria react in the same way or if mitochondria residing in different groups have different sensitivities to calcium. At the moment this is not biologically clarified yet; however, we can introduce such kind of properties into our model by assuming space dependence $f(x, u)$ and $g(x, u)$.

Calcium Evolution

The model consists of spatial developments in terms of diffusing calcium. In addition to the (linear or nonlinear) diffusion term, the equation for the calcium concentration contains a production term dependent on N_2, which is justified by the following:

In the earlier ODE approach of modeling mitochondrial swelling [6] it turned out that it is essential to include the positive feedback mechanism. This accelerating effect is induced by stored Ca^{2+} inside the mitochondria, which is additionally released once the mitochondrion gets completely swollen. Due to a fixed amount of stored calcium, we assume that the additional released calcium is proportional to the newly completely swollen mitochondria, i.e., the mitochondria leaving N_2 and entering N_3. Here the feedback parameter d_2 describes the amount of stored calcium.

In contrast to the previous model, now the action of the positive feedback is contained directly by providing additional calcium. As we will see later on, this additional term is always nonnegative and hence it is interesting to take a look at the long-time behavior of the solution.

References

1. S.V. Baranov, I.G. Stavrovskaya, A.M. Brown, A.M. Tyryshkin, B.S. Kristal, Kinetic model for Ca2+-induced permeability transition in energized liver mitochondria discriminates between inhibitor mechanisms. J. Biol. Chem. **283**(2), 665–676 (2008)
2. S.B. Berman, S.C. Watkins, T.G. Hastings, Quantitative biochemical and ultrastructural comparison of mitochondrial permeability transition in isolated brain and liver mitochondria: evidence for reduced sensitivity of brain mitochondria. Exp. Neurol. **164**(2), 415–425 (2000)
3. P. Bernardi, A. Krauskopf, E. Basso, V. Petronilli, E. Blalchy-Dyson, F. Di Lisa, M.A. Forte, The mitochondrial permeability transition from in vitro artifact to disease target. Febs J. **273**(10), 2077–2099 (2006)
4. N. Brustovetsky, T. Brustovetsky, K.J. Purl, M. Capano, M. Crompton, J.M. Dubinsky, Increased susceptibility of striatal mitochondria to calcium-induced permeability transition. J. Neurosci. **23**(12), 4858–4867 (2003)
5. S. Eisenhofer, B.A. Hense, F. Toókos, H. Zischka, Modeling the volume change in mitochondria. English. Int. J. Biomath. Biostat. **1**(1), 53–62 (2010)
6. S. Eisenhofer, F. Toókos, B.A. Hense, S. Schulz, F. Filbir, H. Zischka, A mathematical model of mitochondrial swelling. BMC Res. Notes **3**(1), 1 (2010)
7. S. Grimm, D. Brdiczka, The permeability transition pore in cell death. Apoptosis **12**(5), 841–855 (2007)
8. F. Ichas, L.S. Jouaville, J.-P Mazat, Mitochondria are excitable organelles capable of generating and conveying electrical and calcium signals. Cell **89**(7), 1145–1153 (1997)
9. J.C. Lagarias, J.A. Reeds, M.H. Wright, P.E. Wright, Convergence properties of the Nelder–Mead simplex method in low dimensions. SIAM J. Optim. **9**(1), 112–147 (1998)
10. S. Massari, Kinetic analysis of the mitochondrial permeability transition. J. Biol. Chem. **271**(50), 31942–31948 (1996)
11. V. Petronilli, C. Cola, S. Massari, R. Colonna, P. Bernardi, Physiological effectors modify voltage sensing by the cyclosporin A-sensitive permeability transition pore of mitochondria. J. Biol. Chem. **268**(29), 21939–21945 (1993)
12. A.V. Pokhilko, F.I. Ataullakhanov, E.L. Holmuhamedov, Mathematical model of mitochondrial ionic homeostasis: three modes of Ca 2+ transport. J. Theor. Biol. **243**(1), 152–169 (2006)
13. V.A. Selivanov, F. Ichas, E.L. Holmuhamedov, L.S. Jouaville, Y.V. Evtodienko, J.-P Mazat, A model of mitochondrial Ca 2+-induced Ca 2+ release simulating the Ca 2+ oscillations and spikes generated by mitochondria. Biophys. Chem. **72**(1), 111–121 (1998)
14. H. Zischka, N. Larochette, F. Hoffmann, D. Hamoller, N. Jagemann, J. Lichtmannegger, L. Jennen, J. Muller-Hocker, F. Roggel, M. Gottlicher et al., Electrophoretic analysis of the mitochondrial outer membrane rupture induced by permeability transition. Anal. Chem. **80**(13), 5051–5058 (2008)

Chapter 4
Mathematical Analysis of Vitro Models

4.1 Neumann Boundary Conditions

As we have elaborated earlier, homogeneous Neumann boundary conditions apply to test tube experiments and hence we are now analyzing the **in vitro model**

$$\partial_t u \;=\; d_1 \Delta_x u + d_2 g(u) N_2 \tag{4.1}$$

$$\partial_t N_1 \;=\; -f(u) N_1 \tag{4.2}$$

$$\partial_t N_2 \;=\; f(u) N_1 - g(u) N_2 \tag{4.3}$$

$$\partial_t N_3 \;=\; g(u) N_2 \tag{4.4}$$

with boundary condition

$$\partial_\nu u \;=\; 0 \quad \text{on } \partial\Omega \tag{4.5}$$

and initial values

$$u(x,0) = u_0(x), \quad N_1(x,0) = N_{1,0}(x), \quad N_2(x,0) = N_{2,0}(x),$$
$$N_3(x,0) = N_{3,0}(x).$$

Remark 4.1 We already characterized the initial functions in (3.5); however, the mathematical analysis can be applied to a more general setting.

© Springer Nature Switzerland AG 2018
M. Efendiev, *Mathematical Modeling of Mitochondrial Swelling*,
https://doi.org/10.1007/978-3-319-99100-9_4

Existence and Uniqueness of Global Solutions

The coupled ODE-PDE model (4.1)–(4.5) describing the in vitro swelling process shall now be analyzed mathematically, thereby following [4, 5]. At first we want to show the well posedness of the model. For that purpose we introduce some assumptions to the model functions f and g.

Condition 4.1 For the model functions $f : \mathbb{R} \to \mathbb{R}$ and $g : \mathbb{R} \to \mathbb{R}$ it holds:

1. Nonnegativity:

$$f(s) \geqslant 0 \quad \forall s \in \mathbb{R}$$
$$g(s) \geqslant 0 \quad \forall s \in \mathbb{R}$$

2. Boundedness:

$$f(s) \leqslant f^* < \infty \quad \forall s \in \mathbb{R}$$
$$g(s) \leqslant g^* < \infty \quad \forall s \in \mathbb{R}$$

   with $f^*, g^* > 0$.
3. Lipschitz continuity:

$$|f(s_1) - f(s_2)| \leqslant L_f |s_1 - s_2| \quad \forall s_1, s_2 \in \mathbb{R}$$
$$|g(s_1) - g(s_2)| \leqslant L_g |s_1 - s_2| \quad \forall s_1, s_2 \in \mathbb{R}$$

   with $L_f, L_g \geqslant 0$.

Remark 4.2 Property (iii) implies bounded derivatives $|f'(s)| \leqslant L_f$ and $|g'(s)| \leqslant L_g$ for all $s \in \mathbb{R}$.

One remarkable characteristic of the model is the following:

If we only take a look at the ODE part and define the total mitochondrial population

$$\overline{N}(x, t) := N_1(x, t) + N_2(x, t) + N_3(x, t),$$

then adding the three Eqs. (4.2)–(4.4), we obtain $\partial_t \overline{N} = 0$. This implies

$$\overline{N}(x, t) = \overline{N}(x) = N_{1,0}(x) + N_{2,0}(x) + N_{3,0}(x) \qquad \forall t \geqslant 0 \quad \forall x \in \Omega, \qquad (4.6)$$

i.e., the total population $\overline{N}$ does not change and is given by the sum of the initial data.

In particular we have

$$\|\overline{N}\|_{L^\infty(\Omega)} \leqslant \|N_{1,0}\|_{L^\infty(\Omega)} + \|N_{2,0}\|_{L^\infty(\Omega)} + \|N_{3,0}\|_{L^\infty(\Omega)} < \infty$$

for initial data from $L^\infty(\Omega)$.

The first aim is to study the model in terms of existence and uniqueness of the solution (u, N_1, N_2, N_3). Here we consider the phase space $H = L^2(\Omega)$.

Theorem 4.1 *Let $\Omega \subset \mathbb{R}^n$ be bounded. Under the assumptions of Condition 4.1 it holds:*

For all initial data $u_0 \in L^2(\Omega)$, $N_{1,0} \in L^\infty(\Omega)$, $N_{2,0} \in L^\infty(\Omega)$ and $N_{3,0} \in L^\infty(\Omega)$ with

$$\|N_{1,0}\|_{L^\infty(\Omega)} + \|N_{2,0}\|_{L^\infty(\Omega)} \neq 0, \tag{4.7}$$

the system (4.1)–(4.5) possesses a unique global solution (u, N_1, N_2, N_3) satisfying

$$u \in C([0, T]; L^2(\Omega))$$

$$\sqrt{t}\, \partial_t u \in L^2(0, T; L^2(\Omega))$$

$$\sqrt{t}\, \Delta_x u \in L^2(0, T; L^2(\Omega))$$

$$N_i \in L^\infty(0, T; L^\infty(\Omega)), \quad i = 1, 2, 3,$$

for all $T > 0$.

Remark 4.3 The additional assumption (4.7) first comes into play in the proof of global existence. However, this condition is not very restrictive, since from

$$\|N_{1,0}\|_{L^\infty(\Omega)} + \|N_{2,0}\|_{L^\infty(\Omega)} = 0$$

it follows

$$N_{1,0} \equiv 0 \quad \text{and} \quad N_{2,0} \equiv 0$$

which leads to the trivial solution $N_1(x, t) = 0$, $N_2(x, t) = 0$, $N_3(x, t) = N_{3,0}(x)$ for all $x \in \Omega$ and the model is reduced to the standard heat equation with homogeneous Neumann boundary conditions.

Proof

1.) Local solution

At first we proof existence and uniqueness of a local solution.

In the following we only take a look at Eqs. (4.1)–(4.3). From the conservation law (4.6) we know that the solution N_3 is then given by the identity

$$N_3(x, t) = \overline{N}(x) - N_1(x, t) - N_2(x, t). \tag{4.8}$$

We work in the Banach space

$$X := C([0, T]; L^2(\Omega)) \text{ with } \|u\|_X = \max_{0 \leqslant t \leqslant T} \|u(t)\|_{L^2(\Omega)}$$

and define the mapping

$$\mathcal{B} : u \in X \mapsto N^u := \begin{pmatrix} N_1^u \\ N_2^u \end{pmatrix} \mapsto \hat{u} = \mathcal{B}(u)$$

in the following way: By N_i^u we denote the solution of the corresponding ODE with respect to a fixed u, i.e., the vector N^u solves the pure ODE system

$$\partial_t N^u = \begin{pmatrix} -f(u)N_1^u \\ f(u)N_1^u - g(u)N_2^u \end{pmatrix} =: F(N^u) \qquad (4.9)$$

$$N^u(x, 0) = \begin{pmatrix} N_{1,0}(x) \\ N_{2,0}(x) \end{pmatrix} . \qquad (4.10)$$

This solution dependent on fixed u is now substituted into the PDE (4.1), which means we are looking for the solution $\hat{u}$ of the pure PDE problem

$$\partial_t \hat{u} = d_1 \Delta_x \hat{u} + d_2 g(\hat{u}) N_2^u$$

$$\hat{u}(x, 0) = u_0(x)$$

$$\partial_\nu \hat{u}|_{\partial\Omega} = 0.$$

The outline of the proof is the following:

1. Show that the ODE system possesses a unique solution N^u for fixed u
2. Show that the PDE corresponding to the solution N^u possesses a unique solution $\hat{u}$
3. Show that the mapping $\mathcal{B}$ possesses a unique fixed-point with $\mathcal{B}(u) = u$.

Step (i): Consider the mapping $u \mapsto N^u$

For every $u \in X$ by Condition 4.1 (ii) we have $f(u) \leqslant f^*$ and $g(u) \leqslant g^*$. We now take a look at the ODE system (4.9), (4.10) and search for the unique solution of this ODE in the product Banach space

$$Y := L^2(\Omega) \times L^2(\Omega) .$$

Obviously it holds $F : Y \to Y$. In order to apply the Piccard-Lindelöf theorem we need Lipschitz-continuity of F:

$$\|F(A) - F(B)\|_Y \overset{!}{\leqslant} L_F \|A - B\|_Y \quad \forall A, B \in Y \text{ with } A = (A_1, A_2), B = (B_1, B_2),$$

$$\|F(A) - F(B)\|_Y^2 = \| - f(u)(A_1 - B_1)\|_{L^2(\Omega)}^2$$

$$+ \|f(u)(A_1 - B_1) - g(u)(A_2 - B_2)\|_{L^2(\Omega)}^2$$

$$\leqslant |f(u)|^2 \|A_1 - B_1\|_{L^2(\Omega)}^2 + 2|f(u)|^2 \|A_1 - B_1\|_{L^2(\Omega)}^2$$

$$+ 2|g(u)|^2 \|A_2 - B_2\|_{L^2(\Omega)}^2$$

$$\leqslant \underbrace{\max\left(3f^{*2}, 2g^{*2}\right)}_{=:L_F^2 \text{ with } L_F > 0} \|A - B\|_Y^2.$$

Thus it follows:

For every $u \in X$ there exists a unique solution $N^u = \begin{pmatrix} N_1^u \\ N_2^u \end{pmatrix} \in C([0, \infty], Y)$.

Our aim now is to obtain upper bounds in $L^\infty(\Omega)$.

Remark 4.4 For general initial data $N_{1,0}(x), N_{2,0}(x)$ we do not have any information about the sign of $N_1^u(x, t)$ and $N_2^u(x, t)$.

For that reason we define for $i = 1, 2$

$$\mathcal{T}^+(x) := \{t \in [0, \infty) : N_i^u(x, t) > 0\} = \bigcup_{j \in J} \mathcal{T}_j^+(x)$$

$$\mathcal{T}^-(x) := \{t \in [0, \infty) : N_i^u(x, t) < 0\} = \bigcup_{k \in K} \mathcal{T}_k^-(x)$$

$$\mathcal{T}^0(x) := \{t \in [0, \infty) : N_i^u(x, t) = 0\} = \bigcup_{l \in L} p_l(x) \cup \mathcal{T}_z(x)$$

where $\mathcal{T}_j^+(x)$, $\mathcal{T}_k^-(x)$, and $\mathcal{T}_z(x)$ are connected intervals and the point $p_l(x)$ is the location of the lth zero. Like that we have the following partition of the time interval $[0, \infty)$ for fixed $x \in \Omega$:

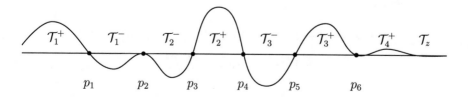

$$[0, \infty) = \mathcal{T}^+(x) \mathbin{\dot\cup} \mathcal{T}^-(x) \mathbin{\dot\cup} \mathcal{T}^0(x).$$

We are now looking at the equations point-wise. For N_1^u it holds

$$\partial_t N_1^u(x, t) = -f(u(x, t))N_1^u(x, t)$$

and that relation immediately implies that if $N_1^u(x, t)$ attains zero at time $t = t_1$, then $N_1^u(x, t) \equiv 0$ for all $t \geq t_1$, i.e., for N_1^u the partition is either given by

$$[0, \infty) = \mathcal{T}^+(x) \mathbin{\dot\cup} \mathcal{T}_z(x) \quad \text{or} \quad [0, \infty) = \mathcal{T}^-(x) \mathbin{\dot\cup} \mathcal{T}_z(x). \tag{4.11}$$

So we only have to consider the cases $N_1^u(x, t) > 0$ for $t \in \mathcal{T}^+(x)$ or $N_1^u(x, t) < 0$ for $t \in \mathcal{T}^-(x)$. For both cases, N_1^u satisfies

$$\partial_t |N_1^u(x, t)| = -f(u(x, t))|N_1^u(x, t)|$$

with $f(u(x, t)) \geq 0$ by Condition 4.1 (i). Hence it easily follows for all $x \in \Omega$ with Gronwall's inequality

$$|N_1^u(x, t)| \leq |N_{1,0}(x)|e^{-f(u(x,t))t} \leq |N_{1,0}(x)|$$

$$\leq \|N_{1,0}\|_{L^\infty(\Omega)} \quad \forall t \in \mathcal{T}^+(x) \text{ or } \forall t \in \mathcal{T}^-(x).$$

For $t \in \mathcal{T}_z(x)$ we clearly have

$$|N_1^u(x, t)| = 0 \leq \|N_{1,0}\|_{L^\infty(\Omega)}$$

and all together by (4.11) we obtain the L^∞ bound

$$\|N_1^u(t)\|_{L^\infty(\Omega)} \leq \|N_{1,0}\|_{L^\infty(\Omega)} =: C_1 < \infty \quad \forall t \geq 0. \tag{4.12}$$

An analogous result can be achieved for N_2^u. However, in that case the partition of $[0, \infty)$ cannot be simplified as before. For fixed $x \in \Omega$ we have to distinguish between different cases.

Case 1: $N_{2,0}(x) \neq 0$, i.e., $p_1(x) \neq 0$

  (a) $N_2^u(x, t) > 0$ on the interval $[0, p_1(x))$, i.e., $[0, p_1(x)) = \mathcal{T}_1^+(x)$
      In that case due to the nonnegativity of g, the boundedness of f and (4.12), it holds

$$\partial_t N_2^u(x, t) = f(u(x, t))N_1^u(x, t) - g(u(x, t))N_2^u(x, t)$$

$$\leq f^* \|N_{1,0}\|_{L^\infty(\Omega)} \quad \forall t \in \mathcal{T}_1^+(x).$$

Hence we easily get by integration $\int_0^t ds$

$$N_2^u(x,t) \leq N_{2,0}(x) + tf^* \|N_{1,0}\|_{L^\infty(\Omega)}$$

$$\leq \|N_{2,0}\|_{L^\infty(\Omega)} + tf^* \|N_{1,0}\|_{L^\infty(\Omega)} \quad \forall t \in \mathcal{T}_1^+(x).$$

(b) $N_2^u(x,t) < 0$ on the interval $[0, p_1(x))$, i.e., $[0, p_1(x)) = \mathcal{T}_1^-(x)$

In contrary to the previous case we have

$$\partial_t N_2^u(x,t) \geq -f(u(x,t))|N_1^u(x,t)| - g(u(x,t))N_2^u(x,t)$$

$$\geq -f^* \|N_{1,0}\|_{L^\infty(\Omega)} \quad \forall t \in \mathcal{T}_1^-(x),$$

which implies

$$N_2^u(x,t) \geq -|N_2(x,0)| - tf^* \|N_{1,0}\|_{L^\infty(\Omega)}$$

$$\geq -\|N_{2,0}\|_{L^\infty(\Omega)} - tf^* \|N_{1,0}\|_{L^\infty(\Omega)} \quad \forall t \in \mathcal{T}_1^-(x).$$

In summary for *Case 1* we obtain

$$|N_2^u(x,t)| \leq \|N_{2,0}\|_{L^\infty(\Omega)} + tf^* \|N_{1,0}\|_{L^\infty(\Omega)} \quad \forall x \in \Omega \quad \forall t \in [0, p_1(x)).$$

Case 2: $N_{2,0}(x) = 0$, i.e., $p_1(x) = 0$

(a) $N_2^u(x,t) > 0$ on the interval $(p_1(x), p_2(x))$, i.e., $(p_1(x), p_2(x)) = \mathcal{T}_1^+(x)$

In analogy to *Case 1* but with $N_2^u(x,0) = N_2^u(x, p_1(x)) = 0$ we obtain by integration $\int_{p_1(x)}^t ds$

$$N_2^u(x,t) \leq (t - p_1(x))f^* \|N_{1,0}\|_{L^\infty(\Omega)}$$

$$\leq \|N_{2,0}\|_{L^\infty(\Omega)} + tf^* \|N_{1,0}\|_{L^\infty(\Omega)} \quad \forall t \in \mathcal{T}_1^+(x).$$

(b) $N_2^u(x,t) < 0$ on the interval $(p_1(x), p_2(x))$, i.e., $(p_1(x), p_2(x)) = \mathcal{T}_1^-(x)$

The same statements yield

$$|N_2^u(x,t)| \leq \|N_{2,0}\|_{L^\infty(\Omega)} + tf^* \|N_{1,0}\|_{L^\infty(\Omega)} \quad \forall t \in \mathcal{T}_1^-(x).$$

With that for the first interval $\mathcal{T}_1^\pm(x)$ we achieved the desired estimate in both cases. Our aim now is to make the transition to the whole time interval $t \geq 0$. It obviously holds

$$|N_2^u(x,t)| = 0 \leq \|N_{2,0}\|_{L^\infty(\Omega)} + tf^* \|N_{1,0}\|_{L^\infty(\Omega)} \quad \forall t \in \mathcal{T}_z(x) \text{ and } \forall t \in \bigcup_{l \in L} p_l(x).$$

By repeating the arguments of *Case 2* to the remaining intervals $[p_l(x), p_{l+1}(x))$ using $t - p_l(x) \leqslant t \; \forall t \in [p_l(x), p_{l+1}(x))$ and $N_2^u(x, p_l(x)) = 0$, with the partition of $[0, \infty)$ we finally obtain

$$\|N_2^u(t)\|_{L^\infty(\Omega)} \leqslant \|N_{2,0}\|_{L^\infty(\Omega)} + tf^*\|N_{1,0}\|_{L^\infty(\Omega)} =: C_2(t) \quad \forall t \geqslant 0. \tag{4.13}$$

Step (ii): Consider the mapping $N^u \longmapsto \hat{u}$

Now we are given the solutions N_1^u and N_2^u for fixed u and take a look at the solution of the corresponding PDE. That means we are focusing on the problem

$$\partial_t \hat{u} = d_1 \Delta_x \hat{u} + d_2 g(\hat{u}) N_2^u$$
$$\hat{u}(x, 0) = u_0(x) \tag{4.14}$$
$$\partial_v \hat{u}|_{\partial\Omega} = 0.$$

In Condition 4.1 (iii) we require the function g to be Lipschitz continuous on $\mathbb{R}$. That can be easily expanded to Lipschitz continuity on $L^2(\Omega)$:

$$\|g(v_1) - g(v_2)\|_{L^2(\Omega)}^2 = \int_\Omega |g(v_1(x)) - g(v_2(x))|^2 \, dx \leqslant L_g^2 \int_\Omega |v_1(x) - v_2(x)|^2 \, dx$$
$$= L_g^2 \|v_1 - v_2\|_{L^2(\Omega)}^2.$$

From that and the properties of the solution N_2^u we are indeed dealing with a parabolic equation with Lipschitz type perturbation. Problems of this type are treated, e.g., in [1, 3] or in a more abstract way in [2, 7]. There the existence and uniqueness of a solution $\hat{u}$ are shown and it is proved that the solution satisfies

$$\hat{u} \in C([0, T]; L^2(\Omega))$$
$$\sqrt{t} \, \partial_t \hat{u} \in L^2(0, T; L^2(\Omega)) \tag{4.15}$$
$$\sqrt{t} \, \Delta_x \hat{u} \in L^2(0, T; L^2(\Omega)).$$

By the property of the solution $\hat{u} \in X$ it follows $\mathcal{B} : X \to X$.
Step (iii): Consider the mapping $u \longmapsto \mathcal{B}(u)$

This mapping maps u to the solution $\hat{u}$ of the PDE problem (4.14), which is dependent on the argument u via the term N_2^u. Our aim now is to find the specific solution $\hat{u} = u$ of (4.14) corresponding to N_2^u. That means we are searching for the fixed point

$$\mathcal{B}(u) = \hat{u} \overset{!}{=} u.$$

If we can show that this fixed point exists and is unique, then we can find the unique solution of the original problem (4.1)–(4.5) and the proof is finished. This is done by the famous contraction mapping theorem, which means we have to show the following:

$\exists\, T_0 > 0$ sufficiently small and $\exists\, \alpha \in (0, 1)$ such that the mapping

$$\mathcal{B} : C([0, T_0]; L^2(\Omega)) =: X_0 \rightarrow X_0$$

is a contraction, i.e.

$$\|\mathcal{B}(u_1) - \mathcal{B}(u_2)\|_{X_0} \leqslant \alpha \|u_1 - u_2\|_{X_0} \quad \forall u_1, u_2 \in X_0.$$

For that we calculate the difference between two solutions of the ODE problem (4.9), (4.10) and the PDE problem (4.14) corresponding to different fixed u_1, $u_2 \in X$:

① $\delta N_1 := N_1^{u_1} - N_1^{u_2}$, where

$$\partial_t N_1^{u_1} = -f(u_1)N_1^{u_1} \qquad\qquad N_1^{u_1}(x, 0) = N_{1,0}(x)$$
$$\partial_t N_1^{u_2} = -f(u_2)N_1^{u_2} \qquad\qquad N_1^{u_2}(x, 0) = N_{1,0}(x).$$

This yields $\delta N_1(x, 0) = 0$ and

$$\begin{aligned}
\partial_t \delta N_1 &= -f(u_1)N_1^{u_1} + f(u_2)N_1^{u_2} \\
&= -f(u_1)N_1^{u_1} + f(u_1)N_1^{u_2} - f(u_1)N_1^{u_2} + f(u_2)N_1^{u_2} \\
&= -f(u_1)\delta N_1 + (f(u_2) - f(u_1))\,N_1^{u_2}.
\end{aligned}$$

Multiplication by δN_1, integration over $\int_\Omega dx$ and using the Lipschitz continuity of f leads to

$$\int_\Omega \partial_t \delta N_1 \delta N_1\, dx = -\int_\Omega f(u_1)|\delta N_1|^2\, dx + \int_\Omega (f(u_2) - f(u_1))\, N_1^{u_2} \delta N_1\, dx$$

$$\Rightarrow \quad \frac{1}{2}\frac{d}{dt}\|\delta N_1\|_{L^2(\Omega)}^2 \leqslant -\int_\Omega f(u_1)|\delta N_1|^2\, dx + L_f \int_\Omega |u_2 - u_1||N_1^{u_2}||\delta N_1|\, dx.$$

With $\delta u := u_1 - u_2$, the nonnegativity of f and the L^∞-estimate (4.12) it follows

$$\frac{1}{2}\frac{d}{dt}\|\delta N_1(t)\|_{L^2(\Omega)}^2 \leqslant L_f C_1 \int_\Omega |\delta u(x, t)||\delta N_1(x, t)|\, dx.$$

and by Hölder's inequality we obtain

$$\frac{d}{dt}\|\delta N_1(t)\|_{L^2(\Omega)}\|\delta N_1(t)\|_{L^2(\Omega)} \leqslant L_f C_1 \|\delta u(t)\|_{L^2(\Omega)}\|\delta N_1(t)\|_{L^2(\Omega)}$$

$$\Rightarrow \quad \frac{d}{dt}\|\delta N_1(t)\|_{L^2(\Omega)} \leqslant L_f C_1 \|\delta u(t)\|_{L^2(\Omega)}.$$

Integrating over $\int_0^t ds$ and using the fact $\delta N_1(0) = 0$, we are led to the final result

$$\|\delta N_1(t)\|_{L^2(\Omega)} \leqslant L_f C_1 \int_0^t \|\delta u(s)\|_{L^2(\Omega)}\, ds \quad \forall t \in [0, T]. \tag{4.16}$$

② $\delta N_2 := N_2^{u_1} - N_2^{u_2}$, where $N_2^{u_1}$, $N_2^{u_2}$ solve

$$\partial_t N_2^{u_1} = f(u_1)N_1^{u_1} - g(u_1)N_2^{u_1} \qquad N_2^{u_1}(x, 0) = N_{2,0}(x)$$
$$\partial_t N_2^{u_2} = f(u_2)N_1^{u_2} - g(u_2)N_2^{u_2} \qquad N_2^{u_2}(x, 0) = N_{2,0}(x).$$

Again we have $\delta N_2(x, 0) = 0$ and

$$\partial_t \delta N_2 = f(u_1)N_1^{u_1} - f(u_2)N_1^{u_2} - g(u_1)N_2^{u_1} + g(u_2)N_2^{u_2}$$
$$= f(u_1)\delta N_1 + (f(u_1) - f(u_2))\, N_1^{u_2} - g(u_1)\delta N_2 + (g(u_2) - g(u_1))\, N_2^{u_2}.$$

Multiplying by δN_2 and integrating over $\int_\Omega dx$ using the boundedness of f and the nonnegativity of g together with the Lipschitz continuity, we get

$$\frac{1}{2}\frac{d}{dt}\|\delta N_2\|_{L^2(\Omega)}^2 \leqslant f^* \int_\Omega |\delta N_1||\delta N_2|\, dx + L_f \int_\Omega |\delta u||N_1^{u_2}||\delta N_2|\, dx$$
$$+ L_g \int_\Omega |\delta u||N_2^{u_2}||\delta N_2|\, dx.$$

We are looking for local solutions and the existence of a time T_0 sufficiently small to assure contraction. Hence without loss of generality we can assume $t \leqslant T \leqslant 1$. Then by (4.13) we have the estimate

$$\|N_2^u(t)\|_{L^\infty(\Omega)} \leqslant C_2(t) \leqslant C_2(1) = \|N_{2,0}\|_{L^\infty(\Omega)} + f^*\|N_{1,0}\|_{L^\infty(\Omega)} =: C_2 < \infty. \tag{4.17}$$

Application of Hölder's inequality, canceling out $\|\delta N_2(t)\|_{L^2(\Omega)}$ and using estimate (4.12) together with the previous result, we get

$$\frac{d}{dt}\|\delta N_2(t)\|_{L^2(\Omega)} \leqslant f^*\|\delta N_1(t)\|_{L^2(\Omega)} + \underbrace{(L_f C_1 + L_g C_2)}_{=:C_3} \|\delta u(t)\|_{L^2(\Omega)}.$$

By substituting estimate (4.16), integrating over $\int_0^t ds$, and using $\delta N_2(0) = 0$ we obtain

$$\|\delta N_2(t)\|_{L^2(\Omega)} \leqslant f^* L_f C_1 \int_0^t \int_0^s \|\delta u(\tau)\|_{L^2(\Omega)}\, d\tau\, ds + C_3 \int_0^t \|\delta u(s)\|_{L^2(\Omega)}\, ds.$$

Due to the fact $t \in [0, T]$ with $T \leqslant 1$ the double integral is estimated by

$$\int_0^t \int_0^s \|\delta u(\tau)\|_{L^2(\Omega)} \, d\tau \, ds \leqslant \int_0^t \int_0^t \|\delta u(\tau)\|_{L^2(\Omega)} \, d\tau \, ds \leqslant \int_0^t \|\delta u(s)\|_{L^2(\Omega)} \, ds \, .$$

This finally gives

$$\|\delta N_2(t)\|_{L^2(\Omega)} \leqslant \underbrace{\left(f^* L_f C_1 + C_3\right)}_{=: C_4} \int_0^t \|\delta u(s)\|_{L^2(\Omega)} \, ds \quad \forall t \in [0, T] \, . \tag{4.18}$$

③ $\delta \hat{u} := \hat{u}_1 - \hat{u}_2 = \mathcal{B}(u_1) - \mathcal{B}(u_2)$, where $\hat{u}_1, \hat{u}_2$ are solutions of

$$\partial_t \hat{u}_1 = d_1 \Delta_x \hat{u}_1 + d_2 g(\hat{u}_1) N_2^{u_1} \qquad \hat{u}_1(x,0) = u_0(x) \qquad \partial_\nu \hat{u}_1|_{\partial\Omega} = 0$$
$$\partial_t \hat{u}_2 = d_1 \Delta_x \hat{u}_2 + d_2 g(\hat{u}_2) N_2^{u_2} \qquad \hat{u}_2(x,0) = u_0(x) \qquad \partial_\nu \hat{u}_2|_{\partial\Omega} = 0 \, .$$

Remark 4.5 Our goal was to show that the mapping $\mathcal{B} : X_0 \to X_0$ is a contraction for an appropriate choice of the time interval $[0, T_0]$. By the definition of δu and $\delta \hat{u}$, $\mathcal{B}$ is a contraction if

$$\max_{0 \leqslant t \leqslant T_0} \|\delta \hat{u}(t)\|_{L^2(\Omega)} \leqslant \alpha \max_{0 \leqslant t \leqslant T_0} \|\delta u(t)\|_{L^2(\Omega)}, \quad 0 < \alpha < 1 \, .$$

It holds $\delta \hat{u}(x, 0) = 0$ and

$$\partial_t \delta \hat{u} = d_1 \Delta_x \delta \hat{u} + d_2 g(\hat{u}_1) N_2^{u_1} - d_2 g(\hat{u}_2) N_2^{u_2}$$
$$= d_1 \Delta_x \delta \hat{u} + d_2 g(\hat{u}_1) \delta N_2 + d_2 \left(g(\hat{u}_1) - g(\hat{u}_2)\right) N_2^{u_2} \, .$$

As before, multiplication by $\delta \hat{u}$, integration over $\int_\Omega dx$, application of Hölder's inequality, and making use of estimate (4.17), $T \leqslant 1$ and the properties of g together with the Neumann boundary condition give

$$\frac{1}{2} \frac{d}{dt} \|\delta \hat{u}(t)\|_{L^2(\Omega)}^2 + d_1 \|\nabla_x \delta \hat{u}(t)\|_{L^2(\Omega)}^2 \leqslant d_2 g^* \|\delta N_2(t)\|_{L^2(\Omega)} \|\delta \hat{u}(t)\|_{L^2(\Omega)}$$
$$+ d_2 L_g C_2 \|\delta \hat{u}(t)\|_{L^2(\Omega)}^2$$

$$\Rightarrow \quad \frac{d}{dt} \|\delta \hat{u}(t)\|_{L^2(\Omega)} \leqslant d_2 g^* \|\delta N_2(t)\|_{L^2(\Omega)}$$
$$+ d_2 L_g C_2 \|\delta \hat{u}(t)\|_{L^2(\Omega)} \, .$$

Substituting the previous result (4.18) and taking $\int_0^t ds$ with $\delta\hat{u}(0) = 0$, we have

$$\|\delta\hat{u}(t)\|_{L^2(\Omega)} \leq d_2 g^* C_4 \int_0^t \int_0^s \|\delta u(\tau)\|_{L^2(\Omega)} d\tau\, ds$$
$$+ d_2 L_g C_2 \int_0^t \|\delta\hat{u}(s)\|_{L^2(\Omega)} ds\,.$$

With the same argument as before, the double integral can be cancelled and we can further estimate by using Gronwall's inequality

$$\|\delta\hat{u}(t)\|_{L^2(\Omega)} \leq d_2 g^* C_4\, T \max_{0 \leq t \leq T} \|\delta u(t)\|_{L^2(\Omega)} + d_2 L_g C_2 \int_0^t \|\delta\hat{u}(s)\|_{L^2(\Omega)} ds$$

$$\Rightarrow \qquad \|\delta\hat{u}(t)\|_{L^2(\Omega)} \leq d_2 g^* C_4\, T \|\delta u\|_X e^{d_2 L_g C_2 T} \qquad \forall t \in [0, T]\,.$$

$$\Rightarrow \qquad \|\delta\hat{u}\|_X \leq \underbrace{d_2 g^* C_4\, T e^{d_2 L_g C_2 T}}_{=: C_5(T) > 0} \|\delta u\|_X\,.$$

From that it follows

$$\mathcal{B} \text{ is a contraction } \Leftrightarrow \exists\, T_0 > 0 \text{ such that } C_5(T_0) < 1\,.$$

This is indeed the case since for $T \to 0$ we have $C_5(T) \to 0$, whence follows the existence of $T_0 = T_0\left(\|N_{1,0}\|_{L^\infty(\Omega)}, \|N_{2,0}\|_{L^\infty(\Omega)}\right)$ with the required properties. Thus we showed

$$\mathcal{B} \text{ is a contraction on the space } X_0 = C([0, T_0], L^2(\Omega))\,.$$

Now the contraction mapping theorem can be applied to $\mathcal{B}$ and yields

$$\exists!\, u \in X_0 : \mathcal{B}(u) = u\,.$$

As noted before, the unique fixed-point u is exactly the solution of the original problem and thus we have the existence and uniqueness of the local solution u. This solution satisfies (4.15) with $T = T_0$.

Furthermore the uniqueness of N_1 and N_2 in $[0, T_0]$ follows from (4.16) and (4.18) immediately, since by the uniqueness of u we have $\delta u = 0$. The conservation identity (4.8) provides the unique solution N_3. For $t \in [0, T_0]$ with $T_0 \leq 1$, estimates (4.12) and (4.17) induce

$$\left.\begin{array}{l} \|N_1(t)\|_{L^\infty(\Omega)} \leq C_1 < \infty \Rightarrow \|N_1\|_{L^\infty(0,T_0;L^\infty(\Omega))} < \infty \\ \|N_2(t)\|_{L^\infty(\Omega)} \leq C_2 < \infty \Rightarrow \|N_2\|_{L^\infty(0,T_0;L^\infty(\Omega))} < \infty \end{array}\right\} \Rightarrow \|N_3\|_{L^\infty(0,T_0;L^\infty(\Omega))} < \infty\,.$$

This completes the proof of local existence and uniqueness.

$$\square\ 1.)$$

2.) Global solution

Our next aim is to show that this local solution exists globally in time, i.e., not only on the time interval $[0, T_0]$ but on any bounded interval $[0, T]$ for all $T > 0$.

For that we come back to the definition of the function $C_5(T)$ determining the time T_0 such that $C_5(T_0) < 1$. For simplification this function can be further estimated by the definition of the appearing constants C_2 and C_4 and the fact $T_0 \leqslant 1$. Here we define

$$s := \|N_{1,0}\|_{L^\infty(\Omega)} + \|N_{2,0}\|_{L^\infty(\Omega)} \qquad > 0 \text{ by (4.7)}$$

$$a_1 := 2d_2 g^* \max\left(L_f + L_g f^* + L_f f^*, L_g\right) > 0$$

$$a_2 := 2d_2 L_g \max\left(1, f^*\right) \qquad\qquad > 0$$

and it follows

$$C_5(T_0) \leqslant a_1 s \, T_0 e^{a_2 s T_0} \leqslant a_1 s \, T_0 e^{a_2 s} \overset{!}{<} 1 \,,$$

i.e., then there exists $0 < \xi < 1$ such that

$$a_1 s \, T_0 e^{a_2 s} = \xi \quad \Leftrightarrow \quad T_0 = \frac{\xi}{a_1 s e^{a_2 s}} =: H(s) \geqslant 0 \,. \tag{4.19}$$

More precisely, the function $H(s)$ is strictly monotonically decreasing due to

$$\frac{d}{ds} H(s) = -\xi \frac{a_1 e^{a_2 s} + a_1 a_2 s e^{a_2 s}}{(a_1 s e^{a_2 s})^2} < 0 \,.$$

Up to this point, we obtained the existence of a unique solution (u, N_1, N_2, N_3) satisfying the properties of Theorem 4.1 on the time interval $[0, T_0]$. Thus we can define new initial conditions

$$u_0(x) = u(x, T_0) \,, \quad N_{1,0}(x) = N_1(x, T_0) \,, \quad N_{2,0}(x) = N_2(x, T_0) \,,$$

$$N_{3,0}(x) = N_3(x, T_0)$$

which satisfy

$$\|N_1(T_0)\|_{L^\infty(\Omega)} + \|N_2(T_0)\|_{L^\infty(\Omega)} \leqslant C_{T_0} < \infty \,.$$

Relation (4.19) implies that we have the existence of a positive time $T_0 > 0$ as long as s is bounded by some constant. This is the case here, whence follows the existence of a time T_1 such that the problems (4.1)–(4.4) with the new initial conditions possess a unique local solution on $[T_0, T_1]$. With that we can extend the solution to the bigger interval $[0, T_1]$.

Hence, in order to assure the existence of the unique global solution, it suffices to show that

$$\sup_{0 \leqslant t \leqslant T} \left(\|N_1(t)\|_{L^\infty(\Omega)} + \|N_2(t)\|_{L^\infty(\Omega)} \right) \leqslant C_T < \infty$$

for any bounded interval $[0, T]$, where the constant C_T is dependent on the time T. If we can show the boundedness of this expression, we can stepwise extend the existence interval of the local solution to any interval $[0, T]$ for all $T > 0$, which means that the solution exists globally in time.

In fact, proceeding in exactly the same way as in Step (i) of the local existence proof to show (4.12) and (4.13), we obtain the following estimates:

$$\|N_1(t)\|_{L^\infty(\Omega)} \leqslant \|N_{1,0}\|_{L^\infty(\Omega)} \qquad\qquad \text{for all } t \geqslant 0,$$

$$\|N_2(t)\|_{L^\infty(\Omega)} \leqslant \|N_{2,0}\|_{L^\infty(\Omega)} + tf^* \|N_{1,0}\|_{L^\infty(\Omega)} \quad \text{for all } t \geqslant 0.$$

This yields the desired result

$$\sup_{0 \leqslant t \leqslant T} \left(\|N_1(t)\|_{L^\infty(\Omega)} + \|N_2(t)\|_{L^\infty(\Omega)} \right)$$

$$\leqslant \|N_{2,0}\|_{L^\infty(\Omega)} + (1 + Tf^*)\|N_{1,0}\|_{L^\infty(\Omega)} =: C_T < \infty$$

which implies that the unique local solution can be continued globally.

$$\Box \, 2.)$$

Theorem 4.2 *Let all assumptions of Theorem 4.1 hold and in addition assume that*

$$u_0 \geqslant 0, \quad N_{1,0} \geqslant 0, \quad N_{2,0} \geqslant 0, \quad N_{3,0} \geqslant 0.$$

Then the solution (u, N_1, N_2, N_3) *preserves nonnegativity. Furthermore* N_1, N_2, *and* N_3 *are uniformly bounded in* $\Omega \times [0, \infty)$.

Proof
1.) Nonnegativity
The property of preserving nonnegativity starting from nonnegative initial values is proved with the standard technique to show that the negative part of the solution is zero. Recall that for a general function v the positive and negative parts are defined by

$$v^+ := \max(v, 0) \geqslant 0, \quad v^- := \max(-v, 0) \geqslant 0 \quad \text{and with that} \quad v = v^+ - v^-.$$

That means for the preservation of nonnegativity we have to show

$$v^-(\cdot, 0) \equiv 0 \quad \Rightarrow \quad v^-(\cdot, t) \equiv 0 \quad \forall t \geqslant 0.$$

We start with proving nonnegativity of unswollen mitochondria.

① $N_1(t) \geqslant 0$

By assumption we have $N_{1,0}(x) \geqslant 0$ and consequently $N_1^-(0) \equiv 0$. Multiplying (4.2)

$$\partial_t N_1 = -f(u)N_1$$

by N_1^-, using $N_1 = N_1^+ - N_1^-$ and integrating over $\int_\Omega dx$, we get with Condition 4.1 (i)

$$\frac{1}{2}\frac{d}{dt}\|N_1^-(t)\|_{L^2(\Omega)}^2 = -\int_\Omega f(u(t))|N_1^-(t)|^2\,dx \leqslant 0,$$

which gives

$$\|N_1^-(t)\|_{L^2(\Omega)}^2 \leqslant \|N_1^-(0)\|_{L^2(\Omega)}^2 = 0.$$

This implies $\|N_1^-(t)\|_{L^2(\Omega)} = 0\ \forall t > 0$, i.e.,

$$N_1^-(t) \equiv 0 \quad \forall t \geqslant 0 \quad \Leftrightarrow \quad N_1(t) \geqslant 0 \quad \forall t \geqslant 0.$$

② $N_2(t) \geqslant 0$

We have $N_2^-(0) \equiv 0$, $N_2 = N_2^+ - N_2^-$ and by the previous result $N_1 = N_1^+$. Multiplying the Eq. (4.3)

$$\partial_t N_2 = f(u)N_1 - g(u)N_2$$

by the negative part N_2^- and taking $\int_\Omega dx$, we have with Condition 4.1 (i)

$$\frac{1}{2}\frac{d}{dt}\|N_2^-(t)\|_{L^2(\Omega)}^2 = -\int_\Omega f(u(t))N_1^+(t)N_2^-(t)\,dx - \int_\Omega g(u(t))|N_2^-(t)|^2\,dx \leqslant 0.$$

As before we obtain

$$N_2^-(t) \equiv 0 \quad \forall t \geqslant 0 \quad \Leftrightarrow \quad N_2(t) \geqslant 0 \quad \forall t \geqslant 0.$$

③ $N_3(t) \geqslant 0$

With the same arguments making use of the result $N_2 = N_2^+$ we get

$$N_3^-(t) \equiv 0 \quad \forall t \geqslant 0 \quad \Leftrightarrow \quad N_3(t) \geqslant 0 \quad \forall t \geqslant 0.$$

④ $u(t) \geqslant 0$

Multiplication of (4.1) by u^- and integration over $\int_\Omega dx$ yields

$$-\frac{1}{2}\frac{d}{dt}\|u^-(t)\|_{L^2(\Omega)}^2 = d_1 \int_\Omega \Delta_x u(t) u^-(t)\,dx + d_2 \int_\Omega g(u(t))N_2(t)u^-(t)\,dx\,.$$

It holds due to Stampacchia's lemma

$$\int_\Omega \Delta_x u(t) u^-(t)\,dx = \int_{\partial\Omega} \nabla_x u(t) u^-(t)\,\vec{n}\,dS$$

$$-\int_\Omega \nabla_x u(t) \nabla_x u^-(t)\,dx = \|\nabla_x u^-(t)\|_{L^2(\Omega)}^2$$

and by using the fact $N_2 = N_2^+$, we obtain

$$\frac{1}{2}\frac{d}{dt}\|u^-(t)\|_{L^2(\Omega)}^2 = -d_1\|\nabla_x u^-(t)\|_{L^2(\Omega)}^2 - d_2 \int_\Omega g(u(t))N_2^+(t)u^-(t)\,dx \leqslant 0\,.$$

This implies in analogy to the cases before

$$u^-(t) \equiv 0 \quad \forall t \geqslant 0 \quad \Leftrightarrow \quad u(t) \geqslant 0 \quad \forall t \geqslant 0\,.$$

2.) Uniform boundedness of (N_1, N_2, N_3)

From the conservation law

$$N_1(x,t) + N_2(x,t) + N_3(x,t) = N_{1,0}(x) + N_{2,0}(x) + N_{3,0}(x) = \overline{N}(x) \in L^\infty(\Omega)$$

for $x \in \Omega$, and the proved nonnegativity of the ODE parts N_1, N_2, and N_3 it follows immediately

$$0 \leqslant N_i(x,t) \leqslant \|\overline{N}\|_{L^\infty(\Omega)} \quad \forall t \geqslant 0, \forall x \in \Omega, \ i = 1,2,3\,. \tag{4.20}$$

Asymptotic Behavior of Solutions

Now the longtime behavior of the solution (u, N_1, N_2, N_3) is studied. This behavior is highly dependent on the special structure of the model functions f and g.

Proposition 4.1 *Let all assumptions of Theorems 4.1 and 4.2 hold and in addition assume $u_0 \neq 0$.*

Then the unique solution u is strictly positive for $t > 0$ and becomes bounded by below:

$$\exists\, t_0 > 0 \text{ and } \exists\, \varrho > 0: \quad u(x,t) \geqslant \varrho > 0 \quad \forall t \geqslant t_0 \quad \forall x \in \Omega\,.$$

Proof The solution u fulfills the PDE problem

$$\partial_t u = d_1 \Delta_x u + d_2 g(u) N_2$$
$$u(x, 0) = u_0(x)$$
$$\partial_\nu u|_{\partial\Omega} = 0.$$

For the proof we introduce a subsolution $\underline{u}$ and show the relation

$$u(x, t) \geqslant \underline{u}(x, t) \geqslant \varrho > 0.$$

First we show that the subsolution $\underline{u}$ is given by the solution of the following auxiliary problem

$$\partial_t \underline{u} = d_1 \Delta_x \underline{u}$$
$$\underline{u}(x, 0) = u_0(x)$$
$$\partial_\nu \underline{u}|_{\partial\Omega} = 0.$$

By Condition 4.1 and Theorem 4.2 we have $g(u) \geqslant 0$, $N_2 \geqslant 0$. Thus it holds

$$\underbrace{\partial_t \underline{u} - d_1 \Delta_x \underline{u}}_{=0} - \underbrace{g(\underline{u}) N_2}_{\geqslant 0} \leqslant \underbrace{\partial_t u - d_1 \Delta_x u - g(u) N_2}_{=0}$$
$$\underline{u}(x, 0) = u(x, 0)$$
$$\partial_\nu \underline{u}|_{\partial\Omega} = \partial_\nu u|_{\partial\Omega}.$$

From the comparison principle it follows

$$\underline{u}(x, t) \leqslant u(x, t) \quad \forall (x, t) \in \Omega_T := \Omega \times (0, T].$$

By $\partial\Omega_T$ we denote the parabolic boundary $\partial\Omega_T = (\{0\} \times \overline{\Omega}) \cup ([0, T] \times \partial\Omega)$. The next step is to show that $\underline{u} > 0$. For that we consider the constant function $\underline{\underline{u}} \equiv 0$, which is a subsolution of the auxiliary problem due to the initial condition $\underline{\underline{u}}(x, 0) = 0$. Thus we have

$$\underline{\underline{u}}(x, t) \leqslant \underline{u}(x, t) \quad \forall (x, t) \in \Omega_T.$$

We assume that there exists $y_0 := (x_0, t_0) \in \Omega_T \backslash \partial\Omega_T$ such that

$$\underline{\underline{u}}(y_0) = \underline{u}(y_0).$$

That means there is an inner point y_0 at which the function $\underline{u}$ attains its maximum. By the strong parabolic maximum principle this implies

$$\underline{u} \equiv \overline{u} \text{ in } \overline{\Omega}_T \text{ ,,}$$

which is a contradiction to $\underline{u}(0) = u_0(x) \not\equiv 0 = \overline{u}(0)$.

It follows: There is no inner point y_0 such that $\overline{u}(y_0) = \underline{u}(y_0) = 0 \Rightarrow \underline{u} > 0$ in $\Omega_T \backslash \partial\Omega_T$.

Furthermore we can also show that $\underline{u}$ is positive on the boundary for all $t > 0$, i.e., on $\partial\Omega_T \backslash (\{0\} \times \Omega)$. We assume that there exists $y_1 := (x_1, t_1)$ with $x_1 \in \partial\Omega$ and $t_1 > 0$ such that

$$\underline{u}(y_1) = 0.$$

Then by the properties of $\underline{u}$, Hopf's maximum principle (see, e.g., [9, 12]) implies

$$\partial_\nu \underline{u}(y_1) \neq 0 \text{ ,,}$$

which is a contradiction to the homogeneous Neumann boundary condition $\partial_\nu \underline{u}(y_1) = 0$.

It follows: There is no point y_1 on the boundary with $\underline{u}(y_1) = 0 \Rightarrow \underline{u} > 0$ in $\partial\Omega_T \backslash (\{0\} \times \Omega)$.

All together it follows:

$$u(x, t) \geq \underline{u}(x, t) > 0 \text{ in } \Omega_T \text{ ,}$$

which means that u is proceeding instantaneously with infinite spreading speed.

It remains to show that u in fact is bounded below by some positive constant. From $u(x, t) > 0 \; \forall (x, t) \in \Omega_T$ it follows

$$\exists \, t_0 > 0 \text{ and } \varrho > 0 \text{ such that } \min_{x \in \Omega} u(x, t_0) \geq \varrho > 0.$$

We define the constant function $\underline{u}_\varrho \equiv \varrho$. Then by the choice of t_0 it holds

$$\partial_t \underline{u}_\varrho - d_1 \Delta_x \underline{u}_\varrho = \partial_t \underline{u} - d_1 \Delta_x \underline{u}$$

$$\underline{u}_\varrho(x, t_0) \leq \underline{u}(x, t_0)$$

$$\partial_\nu \underline{u}_\varrho |_{\partial\Omega} = \partial_\nu \underline{u}|_{\partial\Omega}.$$

Again applying the comparison principle we obtain

$$\underline{u}(x, t) \geq \underline{u}_\varrho \equiv \varrho \qquad \forall t \geq t_0 \quad \forall x \in \Omega.$$

Taking everything together yields the desired result

$$u(x, t) \geqslant \varrho > 0 \qquad \forall t \geqslant t_0 \quad \forall x \in \Omega.$$

$\square$

This result is now used to obtain information about the type of convergence as time goes to infinity. For that we need additional assumptions on the functions f and g.

Condition 4.2 Let the model functions $f : \mathbb{R} \rightarrow \mathbb{R}$ and $g : \mathbb{R} \rightarrow \mathbb{R}$ fulfill Condition 4.1. In addition we assume that there exist constants $C^- \geqslant 0$, $m_1 > 0$, $m_2 > 0$, $\delta > 0$, and $\varrho_0 > 0$ such that the following assertions hold:

1. Starting threshold:

$$f(s) = 0 \quad \forall s \leqslant C^-$$
$$g(s) = 0 \quad \forall s \leqslant 0$$

2. Smoothness in $[C^-, C^- + \delta]$:

$$m_1(s - C^-) \leqslant f'(s) \leqslant m_2(s - C^-) \quad \forall s \in [C^-, C^- + \delta]$$

3. Lower bounds:

$$f(s) \geqslant f(C^- + \delta) > 0 \quad \forall s \geqslant C^- + \delta$$
$$g(s) \geqslant g(\varrho_0) > 0 \qquad \forall s \geqslant \varrho_0 > 0$$

4. Monotonicity in $[0, \varrho_0]$:

$$g'(s) > 0 \quad \forall s \in [0, \varrho_0]$$

In order to show several convergence results we need the following statement:

Proposition 4.2 *Let $y(t)$ and $a(t)$ be nonnegative functions with $y \in C^1([t_0, t_1])$ and $a \in C([t_0, t_1])$ for $0 \leqslant t_0 < t_1 \leqslant \infty$. If the inequality*

$$\frac{d}{dt} y(t) + \gamma y(t) \leqslant a(t) \tag{4.21}$$

is satisfied with $\gamma \geqslant 0$, then for $t_0 \leqslant t \leqslant t_1$ it holds

$$y(t) \leqslant y(t_0) e^{-\gamma(t - t_0)} + \int_{t_0}^{t} a(s) e^{-\gamma(t - s)} \, ds.$$

Depending on the properties of $a(t)$, we can deduce further estimates:

(i) $a(t) \equiv C$ in $[t_0, t_1]$:

$$y(t) \leqslant y(t_0) + \frac{C}{\gamma} \qquad\qquad ;$$

(ii) $\int_0^\infty a(t)\, dt < \infty$:

$$y(t) \leqslant y(t_0)e^{-\gamma(t-t_0)} + \int_{t_0}^\infty a(t)\, dt$$

which implies $y(t) \xrightarrow{t \to \infty} 0$.

Proof Multiplying (4.21) by $e^{\gamma t}$ yields

$$\frac{d}{dt}y(t)e^{\gamma t} + \gamma y(t)e^{\gamma t} = \frac{d}{dt}\left(y(t)e^{\gamma t}\right) \leqslant a(t)e^{\gamma t}$$

and by integration over $\int_{t_0}^t ds$ and multiplication with $e^{-\gamma t}$ we obtain

$$y(t) \leqslant y(t_0)e^{-\gamma(t-t_0)} + \int_{t_0}^t a(s)e^{-\gamma(t-s)}\, ds . \qquad\qquad (4.22)$$

In the first case (i) we can further estimate

$$\leqslant y(t_0)e^{-\gamma(t-t_0)} + \frac{C}{\gamma}\left(1 - e^{-\gamma(t-t_0)}\right)$$

$$\leqslant y(t_0) + \frac{C}{\gamma} .$$

In the second case (ii) the basic estimate (4.22) yields

$$y(t) \leqslant y(t_0)e^{-\gamma(t-t_0)} + \int_{t_0}^t a(s)\, ds \xrightarrow{t \to \infty} \int_{t_0}^\infty a(s)\, ds$$

for all $t_0 \geqslant 0$. From the assumption $\int_0^\infty a(t)\, dt < \infty$ it follows $\int_{t_0}^\infty a(s)\, ds \to 0$ for t_0 sufficiently large which implies $y(t) \xrightarrow{t \to \infty} 0$. $\qquad\qquad \square$

The next theorem gives information about the strong convergence of the solution.

Theorem 4.3 *Let Condition 4.2 hold. Under the assumptions of Theorem 4.2 we have the following strong convergence results:*

$$N_1(t) \xrightarrow{t \to \infty} N_1^\infty \geqslant 0 \qquad in \ L^p(\Omega), \quad 1 \leqslant p < \infty$$

$$N_2(t) \xrightarrow{t \to \infty} N_2^\infty \equiv 0 \qquad in \ L^p(\Omega), \quad 1 \leqslant p < \infty$$

$$N_3(t) \xrightarrow{t \to \infty} N_3^\infty \leqslant \|\overline{N}\|_{L^\infty(\Omega)} \quad in \ L^p(\Omega), \quad 1 \leqslant p < \infty$$

$$u(t) \xrightarrow{t \to \infty} u^\infty \equiv C \qquad in \ L^2(\Omega).$$

Proof

① $N_1(\cdot, t) \to N_1^\infty$

From the model Eq. (4.2), Condition 4.1 (i), and the nonnegativity result it holds in the point-wise sense

$$\partial_t N_1(x, t) = -f(u(x, t)) N_1(x, t) \leqslant 0 \quad \forall t \geqslant 0 \quad \forall x \in \Omega.$$

Thus the sequence is nonincreasing and bounded below by 0 which yields the convergence

$$N_1(x, t) \xrightarrow{t \to \infty} N_1^\infty(x) \geqslant 0 \quad \forall x \in \Omega.$$

From (4.20) we know $\sup_{x \in \Omega} N_1(x, t) \leqslant \|\overline{N}\|_{L^\infty(\Omega)} \ \forall t \geqslant 0$. Since $N_1(x, t)$ is nonincreasing in t it follows

$$\sup_{x \in \Omega} N_1^\infty(x) \leqslant \sup_{x \in \Omega} N_1(x, t) \leqslant \|\overline{N}\|_{L^\infty(\Omega)}. \tag{4.23}$$

Furthermore we have

$$N_1(x, t) = |N_1(x, t)| \leqslant \overline{N}(x) \text{ where } \int_\Omega |\overline{N}(x)| \, dx \leqslant |\Omega| \|\overline{N}\|_{L^\infty(\Omega)} < \infty.$$

Thus the Lebesgue dominated convergence theorem can be applied and yields

$$\lim_{t \to \infty} \int_\Omega |N_1(x, t) - N_1^\infty(x)| \, dx = 0,$$

i.e., we have strong convergence in $L^1(\Omega)$ to the unique limit $N_1^\infty(x)$. This result can be adapted to $L^p(\Omega)$, $1 \leqslant p < \infty$. We note

$$\int_\Omega |N_1(x, t) - N_1^\infty(x)|^p \, dx = \int_\Omega |N_1(x, t) - N_1^\infty(x)|^{p-1} |N_1(x, t) - N_1^\infty(x)| \, dx$$

$$\leqslant \sup_{x \in \Omega} |N_1(x, t) - N_1^\infty(x)|^{p-1} \int_\Omega |N_1(x, t) - N_1^\infty(x)| \, dx.$$

Applying Minkowski's inequality, we get

$$\leqslant 2^{p-2} \left(\sup_{x \in \Omega} |N_1(x,t)|^{p-1} + \sup_{x \in \Omega} |N_1^\infty(x)|^{p-1} \right)$$

$$\int_\Omega |N_1(x,t) - N_1^\infty(x)| \, dx$$

$$\overset{(4.23)}{\leqslant} 2^{p-1} \underbrace{\|\overline{N}\|_{L^\infty(\Omega)}^{p-1}}_{<\infty} \underbrace{\int_\Omega |N_1(x,t) - N_1^\infty(x)| \, dx}_{\longrightarrow 0} \overset{t \to \infty}{\longrightarrow} 0 .$$

This finally yields

$$N_1(\cdot,t) \overset{t \to \infty}{\longrightarrow} N_1^\infty \quad \text{strongly in } L^p(\Omega), \quad 1 \leqslant p < \infty .$$

② $N_3(\cdot,t) \to N_3^\infty$

For $N_3(x,t)$ the model Eq. (4.4) gives

$$\partial_t N_3(x,t) = g\left(u(x,t)\right) N_2(x,t) \geqslant 0 \quad \forall t \geqslant 0 \quad \forall x \in \Omega .$$

Since $N_3(x,t)$ is bounded above by $\|\overline{N}\|_{L^\infty(\Omega)}$, the monotonicity yields almost everywhere convergence

$$N_3(x,t) \overset{t \to \infty}{\longrightarrow} N_3^\infty(x) \leqslant \|\overline{N}\|_{L^\infty(\Omega)} \text{ for a.e. } x \in \Omega .$$

Analogously to ①, $|N_3(x,t)|$ is dominated by $\overline{N}(x)$ and $\sup_{x \in \Omega} N_3(x,t) \leqslant \sup_{x \in \Omega} N_3^\infty(x) \leqslant \|\overline{N}\|_{L^\infty(\Omega)}$. Thus we have the same convergence result

$$N_3(\cdot,t) \overset{t \to \infty}{\longrightarrow} N_3^\infty \quad \text{strongly in } L^p(\Omega), \quad 1 \leqslant p < \infty .$$

③ $N_2(\cdot,t) \to N_2^\infty$

Using (4.6) we obtain the almost everywhere convergence

$$N_2(x,t) = \overline{N}(x) \quad - N_1(x,t) \quad - N_3(x,t)$$
$$\Big\downarrow t \to \infty \quad \Big\downarrow t \to \infty \quad \Big\downarrow t \to \infty$$
$$\overline{N}(x) \quad - N_1^\infty(x) \quad - N_3^\infty(x) \quad =: N_2^\infty(x)$$

and the convergence in $L^p(\Omega)$ follows immediately:

$$\int_\Omega |N_2(x,t) - N_2^\infty(x)|^p \, dx = \int_\Omega |N_1^\infty(x) - N_1(x,t) + N_3^\infty(x) - N_3(x,t)|^p \, dx$$

$$\leqslant 2^{p-1} \Bigg(\underbrace{\int_\Omega |N_1(x,t) - N_1^\infty(x)|^p \, dx}_{\longrightarrow 0} + \underbrace{\int_\Omega |N_3(x,t) - N_3^\infty(x)|^p \, dx}_{\longrightarrow 0} \Bigg) \overset{t \to \infty}{\longrightarrow} 0 .$$

Thus we have

$$N_2(\cdot, t) \xrightarrow{t \to \infty} N_2^\infty \quad \text{strongly in } L^p(\Omega), \quad 1 \leqslant p < \infty. \tag{4.24}$$

④ $u(\cdot, t) \to u^\infty$

In order to show the convergence properties of u we take a look at the eigenvalue problem

$$-\Delta_x \varphi_j(x) = \lambda_j \varphi_j(x), \quad x \in \Omega$$
$$\partial_\nu \varphi_j|_{\partial\Omega} = 0. \tag{4.25}$$

It is well known that this eigenvalue problem with Neumann boundary conditions has the following properties:

1. For the first eigenvalue and the corresponding eigenfunction it holds:

$$\lambda_1 = 0 \quad \text{and} \quad \varphi_1 \equiv C_\varphi.$$

2. The remaining eigenvalues satisfy

$$\lambda_2 > 0 \quad \text{and} \quad \lambda_i \to \infty \text{ as } i \to \infty$$

and the set of eigenfunctions $\{\varphi_i\}_{i \in \mathbb{N}}$, $\|\varphi_j\|_{L^2(\Omega)} = 1$ forms a complete orthonormal system in $L^2(\Omega)$.

Remark 4.6 We denote by $\lambda(\Omega) := |\Omega|$ the Lebesgue measure of the domain Ω. From the normalizing condition $\|\varphi_1\|_{L^2(\Omega)} = 1$ it follows

$$\varphi_1 = C_\varphi = \lambda(\Omega)^{-1/2}. \tag{4.26}$$

The function $u(t) \in L^2(\Omega)$ can be projected onto the space spanned by eigenfunctions:

$$u(x, t) = \sum_{j=1}^\infty a_j(t) \varphi_j(x).$$

For the convergence result we work with the following orthogonal decomposition

$$u(x, t) = a_1(t) \varphi_1(x) + \varphi^\perp(x, t) \tag{4.27}$$

where the orthogonal complement $\varphi^\perp$ of $a_1(t)\varphi_1$ is given by $\varphi^\perp(x, t) = \sum_{j=2}^\infty a_j(t)\varphi_j(x)$ due to the orthogonality of the eigenfunctions. By (4.25) $\varphi^\perp$

also satisfies the zero Neumann boundary condition

$$\partial_\nu \varphi^\perp |_{\partial\Omega} = 0 .$$

The Fourier coefficients are calculated by

$$a_j(t) = \left(u(t), \varphi_j \right)_{L^2(\Omega)}$$

and thus for the first coefficient $a_1(t)$ it holds

$$a_1(t) = \left(u(t), C_\varphi \right)_{L^2(\Omega)} = C_\varphi \int_\Omega u(x,t)\, dx \qquad (4.28)$$

$$\frac{d}{dt} a_1(t) = C_\varphi \int_\Omega \partial_t u(x,t)\, dx .$$

Substituting this relation into PDE (4.1) gives

$$\frac{d}{dt} a_1(t) = C_\varphi \underbrace{\int_\Omega d_1 \Delta_x u(x,t)\, dx}_{=0} + d_2 C_\varphi \int_\Omega \underbrace{g(u(x,t)) N_2(x,t)}_{\geq 0}\, dx \geq 0$$

$$(4.29)$$

since

$$\int_\Omega d_1 \Delta_x u(x,t)\, dx = \int_{\partial\Omega} d_1 \nabla_x u(x,t) \cdot v\, dS - \int_\Omega \nabla_x d_1 \nabla_x u(x,t)\, dx$$

$$= d_1 \int_{\partial\Omega} \partial_\nu u(x,t)\, dS = 0 .$$

Thus the function $a_1(t)$ is nondecreasing. In order to show convergence, we need to find an upper bound. For that we interpret the previous equation in a different way using the relation $g(u(x,t)) N_2(x,t) = \partial_t N_3(x,t)$:

$$\frac{d}{dt} a_1(t) = d_2 C_\varphi \frac{d}{dt} \int_\Omega N_3(x,t)\, dx .$$

Integration of this identity over $\int_0^t ds$ yields the term we have to estimate:

$$a_1(t) = a_1(0) + d_2 C_\varphi \left(\int_\Omega N_3(x,t)\, dx - \int_\Omega N_3(x,0)\, dx \right) .$$

From (4.20) we know that

$$
\begin{aligned}
N_3(x, t) = N_3(x, 0) + \int_0^t \partial_s N_3(x, s)\, ds &= N_{3,0}(x) \\
+ \int_0^t g(u(x, s)) N_2(x, s)\, ds &\leqslant \|\overline{N}\|_{L^\infty(\Omega)}
\end{aligned}
\tag{4.30}
$$

and since $N_{3,0}(x) \geqslant 0$ we obtain

$$
N_3(x, t) - N_{3,0}(x) \leqslant \|\overline{N}\|_{L^\infty(\Omega)}
$$

$$
\Rightarrow \quad \int_\Omega N_3(x, t)\, dx - \int_\Omega N_{3,0}(x)\, dx \leqslant \lambda(\Omega)\|\overline{N}\|_{L^\infty(\Omega)} < \infty.
$$

The constant $a_1(0)$ is given by

$$
a_1(0) = \big(u(0), C_\varphi\big)_{L^2(\Omega)} = C_\varphi \|u_0\|_{L^1(\Omega)} \leqslant C_\varphi \lambda(\Omega)^{1/2} \|u_0\|_{L^2(\Omega)} < \infty.
$$

All together we obtain

$$
a_1(t) \leqslant C_\varphi \lambda(\Omega)^{1/2} \|u_0\|_{L^2(\Omega)} + d_2 C_\varphi \lambda(\Omega) \|\overline{N}\|_{L^\infty(\Omega)} =: C_{a_1} < \infty,
$$

which yields the existence of the limit $a_1^\infty \leqslant C_{a_1}$ such that

$$
a_1(t) \xrightarrow{t \to \infty} a_1^\infty.
\tag{4.31}
$$

In order to show that u converges to a constant function, it is thus sufficient to show that $\varphi^\perp(x, t) \to 0$ as $t \to \infty$ for a.e. $x \in \Omega$.

For this purpose, we need to derive the boundedness of $\|N_2\|_{L^1(0,\infty;L^1(\Omega))}$: From (4.30), Proposition 4.1, and Condition 4.2 (iii), (iv) it holds

$$
\lambda(\Omega)\|\overline{N}\|_{L^\infty(\Omega)} \geqslant \int_\Omega \int_0^t g(u(x, s)) N_2(x, s)\, ds\, dx \geqslant g(\overline{\varrho}) \int_\Omega \int_0^t N_2(x, s)\, ds\, dx,
$$

where $\overline{\varrho} := \min(\varrho, \varrho_0) > 0$.

Permuting the integration order we obtain a nondecreasing sequence

$$
F(t) := \int_0^t \|N_2(s)\|_{L^1(\Omega)}\, ds \leqslant \frac{\lambda(\Omega)}{g(\overline{\varrho})} \|\overline{N}\|_{L^\infty(\Omega)}
$$

which is bounded for every $t \geqslant 0$ and thus converging to $F(\infty) \leqslant \frac{\lambda(\Omega)}{g(\overline{\varrho})} \|\overline{N}\|_{L^\infty(\Omega)}$:

$$\lim_{t \to \infty} F(t) = \int_0^\infty \|N_2(t)\|_{L^1(\Omega)} \, dt = \|N_2\|_{L^1(0,\infty;L^1(\Omega))} \leqslant \frac{\lambda(\Omega)}{g(\overline{\varrho})} \|\overline{N}\|_{L^\infty(\Omega)} \,.$$

$$(4.32)$$

This estimate also gives a characterization of $N_2^\infty(x)$ since the convergence of the integral over the nonnegative integrand $\int_0^\infty \|N_2(t)\|_{L^1(\Omega)} \, dt$ implies the existence of a sequence $\{t_k\}_{k \in \mathbb{N}}$ with $t_k \to \infty$ such that

$$\lim_{k \to \infty} \|N_2(t_k)\|_{L^1(\Omega)} = 0 \,.$$

From (4.24) and the uniqueness of the strong limit it follows

$$\lim_{k \to \infty} \int_\Omega N_2(x, t_k) \, dx = \int_\Omega N_2^\infty(x) \, dx = 0 = \int_\Omega 0 \, dx \,,$$

i.e., $N_2(x, t) \to 0$ in $L^1(\Omega)$ and by the arguments we used earlier we have

$$N_2(x, t) \xrightarrow{t \to \infty} 0 \quad \text{strongly in } L^p(\Omega), \quad 1 \leqslant p < \infty \,.$$

Furthermore $\int_\Omega N_2^\infty(x) \, dx = 0$ implies

$$N_2^\infty(x) \equiv 0 \quad \text{for a.e. } x \in \Omega \,.$$

Now we return to the decomposition of u and our aim is to show

$$a_1(t)\varphi_1 + \varphi^\perp(\cdot, t) = u(\cdot, t) \xrightarrow{t \to \infty} u^\infty \equiv a_1^\infty C_\varphi \quad \text{in } L^2(\Omega) \,.$$

The orthogonal complement $\varphi^\perp$ fulfills by definition

$$\left(\varphi^\perp(t), \varphi_1\right)_{L^2(\Omega)} = C_\varphi \int_\Omega \varphi^\perp(x, t) = 0$$

and thus Wirtinger's inequality (see Sect. 1.2) can be applied to $\varphi^\perp(t)$:

$$\|\varphi^\perp(t)\|_{L^2(\Omega)} \leqslant C_W \|\nabla_x \varphi^\perp(t)\|_{L^2(\Omega)} \,.$$

Substitution of the decomposition into the PDE (4.1) gives

$$\partial_t \left(a_1(t)\varphi_1(x) + \varphi^\perp(x, t)\right) = d_1 \Delta_x \left(a_1(t)\varphi_1(x) + \varphi^\perp(x, t)\right) + d_2 g(u(x, t)) N_2(x, t) \,,$$

which leads to

$$\frac{d}{dt}a_1(t)\varphi_1(x) + \partial_t\varphi^\perp(x,t) = d_1\Delta_x\varphi^\perp(x,t) + d_2 g(u(x,t))N_2(x,t) \quad (4.33)$$

due to the fact $\varphi_1(x) \equiv C_\varphi$. Multiplying that equation by $\varphi^\perp$, integrating over $\int_\Omega dx$, and using the orthogonality, we get

$$\frac{d}{dt}a_1(t)\underbrace{\int_\Omega \varphi_1 \varphi^\perp(x,t)\,dx}_{=0} + \frac{1}{2}\frac{d}{dt}\|\varphi^\perp(t)\|^2_{L^2(\Omega)}$$

$$= -d_1\|\nabla_x\varphi^\perp(t)\|^2_{L^2(\Omega)} + d_2\int_\Omega g(u(x,t))N_2(x,t)\varphi^\perp(x,t)\,dx\,,$$

i.e.

$$\frac{1}{2}\frac{d}{dt}\|\varphi^\perp(t)\|^2_{L^2(\Omega)} + d_1\|\nabla_x\varphi^\perp(t)\|^2_{L^2(\Omega)} = d_2\int_\Omega g(u(x,t))N_2(x,t)\varphi^\perp(x,t)\,dx\,.$$

$$(4.34)$$

Applying Wirtinger's and Hölder's inequality we have

$$\frac{1}{2}\frac{d}{dt}\|\varphi^\perp(t)\|^2_{L^2(\Omega)} + \frac{d_1}{C_W{}^2}\|\varphi^\perp(t)\|^2_{L^2(\Omega)} \leq d_2\|g(u(t))N_2(t)\|_{L^2(\Omega)}\|\varphi^\perp(t)\|_{L^2(\Omega)}$$

and Young's inequality with $\varepsilon = \frac{d_1}{d_2 C_W^2}$ gives

$$\frac{d}{dt}\|\varphi^\perp(t)\|^2_{L^2(\Omega)} + \underbrace{\frac{d_1}{C_W{}^2}}_{=:\gamma > 0}\|\varphi^\perp(t)\|^2_{L^2(\Omega)} \leq \underbrace{\frac{d_2{}^2 C_W{}^2}{d_1}}_{=:C_6 > 0}\|g(u(t))N_2(t)\|^2_{L^2(\Omega)}\,.$$

$$(4.35)$$

Remark 4.7 By the term "applying Young's inequality with ε" we mean that we solve the following scenario:

<u>Aim:</u> For a given expression kAB with numbers $k > 0$, A, $B \geqslant 0$ and a given target constant $C_t > 0$, find a constant $\varepsilon > 0$ such that

$$kAB = k\sqrt{\varepsilon}A\frac{1}{\sqrt{\varepsilon}}B \leqslant k\left(\frac{\varepsilon}{2}A^2 + \frac{1}{2\varepsilon}B^2\right) \stackrel{!}{=} C_t A^2 + C_\varepsilon B^2\,.$$

This implies

$$k\frac{\varepsilon}{2} \stackrel{!}{=} C_t \quad \Leftrightarrow \quad \varepsilon = \frac{2C_t}{k} \quad \Rightarrow \quad C_\varepsilon = \frac{k}{2\varepsilon} = \frac{k^2}{4C_t}.$$

Put

$$y(t) := \|\varphi^\perp(t)\|^2_{L^2(\Omega)},$$

$$a(t) := C_6\|g(u(t))N_2(t)\|^2_{L^2(\Omega)},$$

then our aim is to show $y(t) \to 0$ by use of Proposition 4.2. Here the estimate of $\|N_2\|_{L^1(0,\infty;L^1(\Omega))}$ comes into play: From (4.32) and Condition 4.1 (ii) it follows

$$\int_0^\infty a(t)\,dt = C_6 \int_0^\infty \int_\Omega |g(u(x,t))N_2(x,t)|^2\,dx\,dt$$

$$\leqslant g^{*2}\|\overline{N}\|_{L^\infty(\Omega)}C_6 \int_0^\infty \int_\Omega |N_2(x,t)|\,dx\,dt$$

$$= g^{*2}\|\overline{N}\|_{L^\infty(\Omega)}C_6\|N_2\|_{L^1(0,\infty;L^1(\Omega))}$$

$$\leqslant C_6\lambda(\Omega)\frac{g^{*2}}{g(\overline{\varrho})}\|\overline{N}\|^2_{L^\infty(\Omega)} =: C_7 < \infty. \tag{4.36}$$

Remark 4.8 From that it also follows $\|N_2\|_{L^2(0,\infty;L^2(\Omega))} < \infty$:
By the definition of $\overline{\varrho} > 0$ we know $g(u(x,t)) \geqslant g(\overline{\varrho}) > 0$ which leads to

$$C_6g(\overline{\varrho})^2\|N_2\|^2_{L^2(0,\infty;L^2(\Omega))} \leqslant \int_0^\infty a(t)\,dt \leqslant C_7.$$

Hence the assumptions of Proposition 4.2 are fulfilled and we can conclude

$$y(t) = \|\varphi^\perp(t)\|^2_{L^2(\Omega)} \xrightarrow{t \to \infty} 0.$$

This yields

$$\varphi^\perp(\cdot,t) \xrightarrow{t \to \infty} 0 \quad \text{strongly in } L^2(\Omega).$$

Together with the convergence result (4.31) of $a_1(t)$, this gives the almost everywhere convergence result for $u(x,t)$:

$$u(x,t) = a_1(t)\varphi(x) + \varphi^\perp(x,t)$$
$$\Big\downarrow t \to \infty \qquad \Big\downarrow t \to \infty$$
$$a_1^\infty C_\varphi \qquad + 0 \qquad := u^\infty$$

and we have

$$u(\cdot, t) \xrightarrow{t \to \infty} u^\infty \equiv a_1^\infty C_\varphi \quad \text{strongly in } L^2(\Omega).$$

This completes the proof.

Corollary 4.1 *Under the assumptions of Theorem 4.3 the following additional facts hold:*
We have the convergence

$$\nabla_x \varphi^\perp(\cdot, t) = \nabla_x u(\cdot, t) \xrightarrow{t \to \infty} 0 \text{ in } L^2(\Omega) \tag{4.37}$$

and the following expressions are bounded:

$$\sup_{t > 0} \|u(t)\|_{L^1(\Omega)} < \infty$$

$$\sup_{t > 0} \|u(t)\|_{L^2(\Omega)} < \infty$$

$$\sup_{t \geq \delta} \|\nabla_x u(t)\|_{L^2(\Omega)} < \infty \quad \text{for } \delta > 0$$

$$\sup_{t > 0} \|\varphi^\perp(t)\|_{L^2(\Omega)} < \infty$$

$$\int_0^T \|\nabla_x u(t)\|_{L^2(\Omega)}^2 \, dt < \infty.$$

Furthermore the following integrals converge, where (4.39) only holds if we make the additional assumption $u_0 \in H^1(\Omega)$

$$\int_0^\infty \|\nabla_x \varphi^\perp(t)\|_{L^2(\Omega)}^2 \, dt < \infty \tag{4.38}$$

$$\int_0^\infty \|\Delta_x \varphi^\perp(t)\|_{L^2(\Omega)}^2 \, dt < \infty \text{ if } u_0 \in H^1(\Omega). \tag{4.39}$$

Proof The first result is obtained in analogy to the previous case $\varphi^\perp(x, t) \to 0$. For that we use again equation (4.33) in terms of the decomposition of u, multiply by $-\Delta_x \varphi^\perp$ and integrate over $\int_\Omega dx$. From the zero Neumann boundary condition it follows

$$(-\Delta_x \varphi^\perp(t), \varphi_1)_{L^2(\Omega)} = (-\Delta_x \varphi^\perp(t), C_\varphi)_{L^2(\Omega)} = 0$$

and we have

$$\frac{1}{2}\frac{d}{dt}\|\nabla_x\varphi^\perp(t)\|^2_{L^2(\Omega)} + d_1\|\Delta_x\varphi^\perp(t)\|^2_{L^2(\Omega)}$$

$$= -d_2\int_\Omega g(u(x,t))N_2(x,t)\Delta_x\varphi^\perp(x,t)\,dx$$

$$\leqslant d_2\|g(u(t))N_2(t)\|_{L^2(\Omega)}\|\Delta_x\varphi^\perp(t)\|_{L^2(\Omega)}.$$

Young's inequality with $\varepsilon = \frac{d_1}{d_2}$ yields

$$\frac{d}{dt}\|\nabla_x\varphi^\perp(t)\|^2_{L^2(\Omega)} + d_1\|\Delta_x\varphi^\perp(t)\|^2_{L^2(\Omega)} \leqslant \frac{d_2{}^2}{d_1}\|g(u(t))N_2(t)\|^2_{L^2(\Omega)} \qquad (4.40)$$

and by use of Corollary 1.1 we get

$$\frac{d}{dt}\|\nabla_x\varphi^\perp(t)\|^2_{L^2(\Omega)} + \frac{d_1}{C_W{}^2}\|\nabla_x\varphi^\perp(t)\|^2_{L^2(\Omega)} \leqslant \frac{d_2{}^2}{d_1}\|g(u(t))N_2(t)\|^2_{L^2(\Omega)}.$$
$$(4.41)$$

As before, we set

$$y(t) := \quad \|\nabla_x\varphi^\perp(t)\|^2_{L^2(\Omega)}$$

$$a(t) := \frac{d_2{}^2}{d_1}\|g(u(t))N_2(t)\|^2_{L^2(\Omega)}.$$

From the differential inequality

$$\frac{d}{dt}y(t) + \gamma y(t) \leqslant a(t)$$

with the same constant $\gamma = \frac{d_1}{C_W{}^2} > 0$ and the same arguments as in the previous case $\varphi^\perp(x,t) \to 0$, by Proposition 4.2 it follows

$$y(t) = \|\nabla_x\varphi^\perp(t)\|^2_{L^2(\Omega)} \xrightarrow{t\to\infty} 0.$$

This yields

$$\nabla_x\varphi^\perp(\cdot,t) \xrightarrow{t\to\infty} 0 \quad \text{strongly in } L^2(\Omega).$$

Now we want to analyze the solution u in terms of the supremum norm over the time interval $(0, \infty]$. Integrating the PDE

$$\partial_t u = d_1\Delta_x u + d_2 g(u)N_2 \qquad (4.42)$$

over $\int_\Omega dx$, we obtain due to the boundary condition

$$\frac{d}{dt}\|u(t)\|_{L^1(\Omega)} = d_2 \int_\Omega g(u(x,t))N_2(x,t)\,dx = d_2\frac{d}{dt}\|N_3(t)\|_{L^1(\Omega)} \geqslant 0$$

and thus the sequence $\|u(t)\|_{L^1(\Omega)}$ is monotonically increasing. Furthermore it is bounded for every $t > 0$, which is shown by integration over $\int_0^t ds$:

$$\|u(t)\|_{L^1(\Omega)} = \|u_0\|_{L^1(\Omega)} - d_2\|N_{3,0}\|_{L^\infty(\Omega)} + d_2\|N_3(t)\|_{L^1(\Omega)}$$

$$\leqslant \|u_0\|_{L^1(\Omega)} - d_2\|N_{3,0}\|_{L^\infty(\Omega)} + \lambda(\Omega)\|\overline{N}\|_{L^\infty(\Omega)} =: C_8 < \infty.$$

This yields

$$\sup_{t>0}\|u(t)\|_{L^1(\Omega)} \leqslant C_8.$$

This result is now used to give an upper bound for $\|u(t)\|_{L^2(\Omega)}$. Multiplying (4.42) by u and integrating over $\int_\Omega dx$, we obtain

$$\frac{1}{2}\frac{d}{dt}\|u(t)\|^2_{L^2(\Omega)} + d_1\|\nabla_x u(t)\|^2_{L^2(\Omega)} \leqslant d_2 g^*\|\overline{N}\|_{L^\infty(\Omega)}\|u(t)\|_{L^1(\Omega)}$$

$$\leqslant d_2 g^*\|\overline{N}\|_{L^\infty(\Omega)}\sup_{t>0}\|u(t)\|_{L^1(\Omega)} =: C_9 < \infty.$$

$$(4.43)$$

By the nonnegativity of the norm we can deduce $\frac{d}{dt}\|u(t)\|_{L^2(\Omega)} < \infty$. Together with the property $u \in C([0,T]; L^2(\Omega))$ and the convergence $\|u(t)\|_{L^2(\Omega)} \to \|u^\infty\|_{L^2(\Omega)}$, this gives

$$\sup_{t>0}\|u(t)\|_{L^2(\Omega)} < \infty.$$

In order to obtain a similar result for $\nabla_x u$, we first have to show the boundedness of $\int_0^T \|\nabla_x u(t)\|^2_{L^2(\Omega)}\,dt$. Integrating estimate (4.43) over $\int_0^T dt$, we get

$$\frac{1}{2}\|u(T)\|^2_{L^2(\Omega)} + d_1\int_0^T \|\nabla_x u(t)\|^2_{L^2(\Omega)}\,dt \leqslant \frac{1}{2}\|u_0\|^2_{L^2(\Omega)} + C_9 T =: C_{10},$$

which is a bounded expression due to the initial condition $u_0 \in L^2(\Omega)$ and yields

$$\int_0^T \|\nabla_x u(t)\|^2_{L^2(\Omega)}\,dt \leqslant \frac{1}{d_1}C_{10}. \qquad (4.44)$$

By the orthogonal decomposition (4.27) with $\varphi_1 \equiv C_\varphi$, we know $\nabla_x \varphi^\perp = \nabla_x u$ and thus (4.41) also reads with Condition 4.1 (ii) and (4.20)

$$\frac{d}{dt}\|\nabla_x u(t)\|^2_{L^2(\Omega)} + \frac{d_1}{C_W^2}\|\nabla_x u(t)\|^2_{L^2(\Omega)} \leq \frac{d_2^2}{d_1}g^{*2}\|\overline{N}\|^2_{L^\infty(\Omega)}\lambda(\Omega) =: C_{11}.$$

Again we set $y(t) := \|\nabla_x u(t)\|^2_{L^2(\Omega)}$ and apply Proposition 4.2 with $\gamma = \frac{d_1}{C_W^2}$ and constant right-hand side $C = C_{11}$. This gives

$$\|\nabla_x u(t)\|^2_{L^2(\Omega)} \leq \|\nabla_x u(\delta)\|^2_{L^2(\Omega)} + \frac{C_{11}}{\gamma} \quad \text{for } t \geq \delta, \tag{4.45}$$

i.e., for the boundedness we need $\|\nabla_x u(\delta)\|_{L^2(\Omega)} < \infty$.

At this we know $\delta > 0$ since we only have initial data from $L^2(\Omega)$. However, from estimate (4.44) we can deduce the existence of a time $\delta \in (0, T]$ such that $\|\nabla_x u(\delta)\|_{L^2(\Omega)} < \infty$ and thus it follows for a suitable constant $C_{12} < \infty$

$$\sup_{t \geq \delta} \|\nabla_x u(t)\|_{L^2(\Omega)} \leq C_{12}. \tag{4.46}$$

Remark 4.9

1) Since $\nabla_x u = \nabla_x \varphi^\perp$, we also have

$$\sup_{t \geq \delta} \|\nabla_x \varphi^\perp(t)\|_{L^2(\Omega)} \leq C_{12}.$$

2) If we assume $u_0 \in H^1(\Omega)$, then (4.45) also holds for $\delta = 0$, i.e., in that case we have

$$\sup_{t > 0} \|\nabla_x u(t)\|_{L^2(\Omega)} \leq C_{13} \tag{4.47}$$

with $C_{13} := \left(\|\nabla_x u_0\|^2_{L^2(\Omega)} + \frac{C_{11}}{\gamma}\right)^{\frac{1}{2}} < \infty$.

Our next aim is to estimate $\sup_{t>0} \|\varphi^\perp(t)\|_{L^2(\Omega)}$. This is done by use of inequality (4.35), which yields after integration $\int_0^t ds$

$$\|\varphi^\perp(t)\|^2_{L^2(\Omega)} \leq \|\varphi^\perp(0)\|^2_{L^2(\Omega)} + C_6 \int_0^t \|g(u(s))N_2(s)\|^2_{L^2(\Omega)} ds$$

$$\leq \|\varphi^\perp(0)\|^2_{L^2(\Omega)} + C_6 \int_0^\infty \|g(u(s))N_2(s)\|^2_{L^2(\Omega)} ds.$$

From (4.36) we know that the integral is bounded and thus we only need $\varphi^\perp(0) \in L^2(\Omega)$. This follows from the orthogonal representation of $u_0 \in L^2(\Omega)$

$$u_0 = \sum_{j=1}^{\infty} a_j(0)\varphi_j = \sum_{j=1}^{\infty} (u_0, \varphi_j)_{L^2(\Omega)}\varphi_j$$

$$\Rightarrow \quad \varphi^\perp(0) = \sum_{j=2}^{\infty} (u_0, \varphi_j)_{L^2(\Omega)}\varphi_j = u_0 - (u_0, C_\varphi)_{L^2(\Omega)}C_\varphi$$

$$\Rightarrow \quad \|\varphi^\perp(0)\|_{L^2(\Omega)} \leqslant \|u_0\|_{L^2(\Omega)} + |(u_0, C_\varphi)_{L^2(\Omega)}|\|C_\varphi\|_{L^2(\Omega)}$$

$$\leqslant \|u_0\|_{L^2(\Omega)}^2 + \|u_0\|_{L^2(\Omega)}\underbrace{\|C_\varphi\|_{L^2(\Omega)}}_{=1}\|C_\varphi\|_{L^2(\Omega)} = 2\|u_0\|_{L^2(\Omega)} < \infty.$$

$$(4.48)$$

In summary this gives

$$\sup_{t>0} \|\varphi^\perp(t)\|_{L^2(\Omega)} < \infty.$$

The next estimate is obtained from inequality (4.34). Application of Hölder's, Wirtinger's, and Young's inequality with $\varepsilon = \frac{d_1}{d_2 C_W}$ yields

$$\frac{1}{2}\frac{d}{dt}\|\varphi^\perp(t)\|_{L^2(\Omega)}^2 + \frac{d_1}{2}\|\nabla_x\varphi^\perp(t)\|_{L^2(\Omega)}^2 \leqslant \frac{d_2^2 C_W^2}{2d_1}\|g(u(t))N_2(t)\|_{L^2(\Omega)}^2.$$

Integration over $\int_0^t ds$ and omitting the positive term $\|\varphi^\perp(t)\|_{L^2(\Omega)}^2$ give

$$d_1 \int_0^t \|\nabla_x\varphi^\perp(s)\|_{L^2(\Omega)}^2 ds \leqslant \|\varphi^\perp(0)\|_{L^2(\Omega)}^2 + \frac{d_2^2 C_W^2}{d_1}\int_0^t \|g(u(s))N_2(s)\|_{L^2(\Omega)}^2 ds.$$

This holds true for every $t > 0$, whence follows with (4.36) and (4.48)

$$\int_0^\infty \|\nabla_x\varphi^\perp(t)\|_{L^2(\Omega)}^2 dt \leqslant \frac{1}{d_1}\|\varphi^\perp(0)\|_{L^2(\Omega)}^2 + \frac{d_2^2 C_W^2}{d_1^2}\int_0^\infty \|g(u(t))N_2(t)\|_{L^2(\Omega)}^2 dt$$

$$\leqslant \frac{2}{d_1}\|u_0\|_{L^2(\Omega)}^2 + \frac{d_2^2 C_W^2}{d_1^2}\lambda(\Omega)\frac{g^{*2}}{g(\varrho)}\|\overline{N}\|_{L^\infty(\Omega)}^2 =: C_{14} < \infty.$$

Thus we have

$$\int_0^\infty \|\nabla_x\varphi^\perp(t)\|_{L^2(\Omega)}^2 dt \leqslant C_{14}.$$

In an analogous way we now want to show the same result for $\Delta_x \varphi^\perp$. Here we use inequality (4.40) and integrate over $\int_0^t ds$. Again omitting the positive term $\|\nabla_x \varphi^\perp(t)\|^2_{L^2(\Omega)}$ and passing to the limit $t \to \infty$ we obtain

$$\int_0^\infty \|\Delta_x \varphi^\perp(t)\|^2_{L^2(\Omega)} \, dt \leqslant \frac{1}{d_1} \|\nabla_x \varphi^\perp(0)\|^2_{L^2(\Omega)} + \frac{d_2^2}{d_1^2} \int_0^\infty \|g(u(t)) N_2(t)\|^2_{L^2(\Omega)} \, dt \, .$$

Remark 4.10 At this point we explicitly need $\nabla_x \varphi^\perp(0) \in L^2(\Omega)$. For all the observations before it was sufficient to have $\varphi^\perp(0) \in L^2(\Omega)$, which follows immediately from $u_0 \in L^2(\Omega)$ as it was shown in (4.48). In order to have $\nabla_x \varphi^\perp(0) \in L^2(\Omega)$, we need to assume $u_0 \in H^1(\Omega)$. By the definition of $\varphi^\perp$ it holds

$$\nabla_x \varphi^\perp(0) = \sum_{j=2}^\infty (u_0, \varphi_j) \nabla_x \varphi_j = \nabla_x u_0 - (u_0, \varphi_1)_{L^2(\Omega)} \nabla_x \varphi_1 = \nabla_x u_0 \, .$$

If $u_0 \in H^1(\Omega)$, then we obtain by (4.36)

$$\int_0^\infty \|\Delta_x \varphi^\perp(t)\|^2_{L^2(\Omega)} \, dt \leqslant \frac{1}{d_1} \|\nabla_x u_0\|^2_{L^2(\Omega)} + \frac{d_2^2}{d_1^2} \lambda(\Omega) \frac{g^{*2}}{g(\varrho)} \|\overline{N}\|^2_{L^\infty(\Omega)} =: C_{15} < \infty \, ,$$

$$\int_0^\infty \|\Delta_x \varphi^\perp(t)\|^2_{L^2(\Omega)} \, dt \leqslant C_{15} \, .$$

□

4.2   Classification of Partial and Complete Swelling

The mitochondrial swelling process and its extent are dependent on the local calcium dose. If the initial concentration u_0 stays below the initiation threshold C^- at any point $x \in \Omega$, then no swelling will happen and we have $N_i(x, t) \equiv N_{i,0}(x)$ $\forall x \in \Omega$, $i = 1, 2, 3$.

Another possible scenario is the so-called "partial swelling." This effect of partial swelling occurs in the experiments and can also be seen in the simulations when the initial calcium concentration lies above C^- at a small region but due to diffusion it does not stay above this threshold for the whole time. This leads to $N_1(x, t) = N_1(x, T_1) \; \forall t \geqslant T_1$ and $N_3(x, t) = N_3(x, T_2) \; \forall t \geqslant T_2$.

But if the initial calcium distribution together with the influence of the positive feedback is sufficiently high, then "complete swelling" occurs which means $N_1(x, t) \to 0$ and $N_3(x, t) \to \overline{N}(x)$ for all $x \in \Omega$.

As it was shown before, for both cases it holds $N_2(x, t) \to 0$.

Condition 4.3 Let the assumption of Conditions 4.1 and 4.2 holds. In addition we assume more regularity of the initial data:

$$u_0 \in H^1(\Omega)$$

$$N_{1,0} \in H^1(\Omega)$$

$$N_{2,0} \in H^1(\Omega)$$

A crucial point to distinguish between partial and complete swelling is to check if $f(u)$ stays positive for all times. For that it is necessary to have uniform convergence of $u(x, t)$ to $u^\infty \equiv a_1^\infty C_\varphi$. Up to now we only have strong convergence in $L^2(\Omega)$. So our aim now is to show uniform convergence, which turns out to be an extensive task.

Theorem 4.4 *Under the assumptions of Condition 4.3, the following additional statements hold:*

$$\sup_{t>0} \|\nabla_x N_1(t)\|_{L^2(\Omega)} < \infty$$

$$\sup_{t>0} \|\nabla_x N_2(t)\|_{L^2(\Omega)} < \infty$$

$$\sup_{t \geqslant \tilde{\delta}} \|\Delta_x \varphi^\perp(t)\|_{L^2(\Omega)} < \infty \quad \text{for } \tilde{\delta} > 0.$$

Furthermore we have uniform convergence

$$\|u(t) - u^\infty\|_{L^\infty(\Omega)} \longrightarrow 0 \quad \text{as } t \to \infty.$$

Proof The course of action is to show the implications

$$\sup_{t>0} \|\nabla_x N_2(t)\|_{L^2(\Omega)} \leqslant C \implies \sup_{t \geqslant \tilde{\delta}} \|\Delta_x \varphi^\perp(t)\|_{L^2(\Omega)} \leqslant C$$

$$\implies \|u(x, t) - u^\infty\|_{C^\alpha(\Omega)} \xrightarrow{t \to \infty} 0$$

for $t \geqslant \tilde{\delta}$, where $C^\alpha(\Omega)$ denotes the Hölder space $C^{0,\alpha}(\Omega)$, $\alpha \in (0, 1]$.
    For that we need the following statement:

Lemma 4.1 ([11]) *Let $v \in C^\alpha(\Omega)$. Then there exist positive constants $\theta > 0$ and $C_\theta > 0$ such that*

$$\|v\|_{C^\alpha(\Omega)} \leqslant C_\theta \|\nabla_x v\|_{L^2(\Omega)}^\theta \|\Delta_x v\|_{L^2(\Omega)}^{1-\theta}.$$

For our problem the domain of $-\Delta_x$ is given by $D(-\Delta_x) = \{v \in H^2(\Omega) : \partial_\nu v|_{\partial\Omega} = 0\}$ and thus by the previous result with $k = 2$ the solution $u(t)$ as well as the constant function u^∞ lie in $C^\alpha(\Omega)$. So we can apply Lemma 4.1 to the difference $u(t) - u^\infty$, which yields by Lemma 4.1 for $t \geqslant \tilde{\delta}$

$$\|u(t) - u^\infty\|_{L^\infty(\Omega)} \leqslant C\|u(t) - u^\infty\|_{C^\alpha(\Omega)} \leqslant CC_\theta \|\nabla_x(u(t) - u^\infty)\|^\theta_{L^2(\Omega)}$$

$$\|\Delta_x(u(t) - u^\infty)\|^{1-\theta}_{L^2(\Omega)}$$

$$= CC_\theta \|\nabla_x \varphi^\perp(t)\|^\theta_{L^2(\Omega)} \|\Delta_x \varphi^\perp(t)\|^{1-\theta}_{L^2(\Omega)}$$

$$\leqslant CC_\theta \sup_{t \geqslant \tilde{\delta}} \|\Delta_x \varphi^\perp(t)\|^{1-\theta}_{L^2(\Omega)} \|\nabla_x \varphi^\perp(t)\|^\theta_{L^2(\Omega)} \,.$$

From (4.37) we know $\|\nabla_x \varphi^\perp(t)\|_{L^2(\Omega)} \xrightarrow{t \to \infty} 0$ and so we conclude

$$\|u(t) - u^\infty\|_{L^\infty(\Omega)} \xrightarrow{t \to \infty} 0 \text{ if } \sup_{t \geqslant \tilde{\delta}} \|\Delta_x \varphi^\perp(t)\|_{L^2(\Omega)} \leqslant C \,,$$

i.e., in that case we have uniform convergence of $u(x, t)$ to u^∞ for $t \geqslant \tilde{\delta}$.

Remark 4.11 The restriction $t \geqslant \tilde{\delta} > 0$ does not pose any problems. We are interested in the longtime dynamics, where we study the behavior for $t \to \infty$ and the small initial interval $[0, \tilde{\delta})$ need not be taken into account.

So our aim now is to show that $\|\Delta_x \varphi^\perp(t)\|_{L^2(\Omega)}$ is bounded for all $t \geqslant \tilde{\delta}$. For that purpose we introduce the operator $A = -\Delta_x$ to be the Laplacian with $D(A) = \{v \in H^2(\Omega) : \partial_\nu v|_{\partial\Omega} = 0\}$, which is well known to be positive and self-adjoint. Thus we can define its square root $A^{\frac{1}{2}}$ which inherits the property of self-adjointness. The norm $\|A^{\frac{1}{2}} v\|_{L^2(\Omega)}$ is then given by

$$\|A^{\frac{1}{2}} v\|^2_{L^2(\Omega)} = (A^{\frac{1}{2}} v, A^{\frac{1}{2}} v)_{L^2(\Omega)} = (v, Av)_{L^2(\Omega)} = (\nabla_x v, \nabla_x v)_{L^2(\Omega)} = \|\nabla_x v\|^2_{L^2(\Omega)} \,.$$

PDE (4.1) is written in terms of A

$$\partial_t u + d_1 Au = d_2 g(u) N_2$$

and the application of $A^{\frac{1}{2}}$ to this equation gives us

$$\partial_t A^{\frac{1}{2}} u + d_1 A^{\frac{3}{2}} u = d_2 A^{\frac{1}{2}} (g(u) N_2) \,.$$

Multiplying by $A^{\frac{3}{2}} u$ and integrating over $\int_\Omega dx$ we obtain due to $A^{\frac{1}{2}}$ being self-adjoint

$$\frac{1}{2} \frac{d}{dt} \|Au(t)\|^2_{L^2(\Omega)} + d_1 \|A^{\frac{3}{2}} u(t)\|^2_{L^2(\Omega)} = d_2 \left(A^{\frac{1}{2}} (g(u(t)) N_2(t)), A^{\frac{3}{2}} u(t) \right)_{L^2(\Omega)} \,.$$

This term can be further estimated by Cauchy Schwarz and Young's inequality with $\varepsilon = \frac{d_1}{d_2}$, whence it follows

$$\frac{1}{2}\frac{d}{dt}\|Au(t)\|_{L^2(\Omega)}^2 + \frac{d_1}{2}\|A^{\frac{3}{2}}u(t)\|_{L^2(\Omega)}^2 \leqslant \frac{d_2^2}{2d_1}\|A^{\frac{1}{2}}\left(g(u(t))N_2(t)\right)\|_{L^2(\Omega)}^2$$

and from the norm definition we have

$$\frac{1}{2}\frac{d}{dt}\|\Delta_x u(t)\|_{L^2(\Omega)}^2 + \frac{d_1}{2}\|A^{\frac{3}{2}}u(t)\|_{L^2(\Omega)}^2 \leqslant \frac{d_2^2}{2d_1}\|\nabla_x\left(g(u(t))N_2(t)\right)\|_{L^2(\Omega)}^2$$

$$\leqslant \frac{d_2^2}{d_1}\|g'(u(t))\nabla_x u(t)N_2(t)\|_{L^2(\Omega)}^2 + \frac{d_2^2}{d_1}\|g(u(t))\nabla_x N_2(t)\|_{L^2(\Omega)}^2.$$

Next we would like to estimate the term $\left\|A^{\frac{3}{2}}u\right\|_{L^2(\Omega)}$ via $\|Au\|_{L^2(\Omega)}$. Indeed, from the decomposition of $u(x,t)$ it follows that

$$Au = A\varphi^\perp = -\Delta_x\varphi^\perp, \quad \partial_\nu u|_{\partial\Omega} = \partial_\nu\varphi^\perp|_{\partial\Omega} = 0 \quad \text{and} \quad \int_\Omega \varphi^\perp\,dx = 0.$$
$$(4.49)$$

From (4.49) it follows that

$$\left\|\nabla_x\varphi^\perp\right\|_{L^2(\Omega)}^2 = \left(-\Delta_x\varphi^\perp, \varphi^\perp\right)_{L^2(\Omega)} \leqslant \left\|\Delta_x\varphi^\perp\right\|_{L^2(\Omega)}\left\|\varphi^\perp\right\|_{L^2(\Omega)}$$
$$\leqslant C_w\left\|\Delta_x\varphi^\perp\right\|_{L^2(\Omega)}\left\|\nabla_x\varphi^\perp\right\|_{L^2(\Omega)}. \qquad (4.50)$$

Estimate (4.50) yields that

$$\|\nabla_x u\|_{L^2(\Omega)} = \left\|\nabla_x\varphi^\perp\right\|_{L^2(\Omega)} \leqslant C_w\left\|\Delta_x\varphi^\perp\right\|_{L^2(\Omega)} = C_w\|\Delta_x u\|_{L^2(\Omega)}. \qquad (4.51)$$

On the other hand, taking into account (4.51) we have that

$$\left\|-\Delta_x\varphi^\perp\right\|_{L^2(\Omega)}^2 = \left(-\Delta_x\varphi^\perp, -\Delta_x\varphi^\perp\right)_{L^2(\Omega)} = \left(A\varphi^\perp, A\varphi^\perp\right)_{L^2(\Omega)}$$
$$= \left(A^{\frac{1}{2}}\varphi^\perp, A^{\frac{3}{2}}\varphi^\perp\right)_{L^2(\Omega)}$$
$$\leqslant \left\|A^{\frac{1}{2}}\varphi^\perp\right\|_{L^2(\Omega)}\left\|A^{\frac{3}{2}}\varphi^\perp\right\|_{L^2(\Omega)} = \left\|\nabla_x\varphi^\perp\right\|_{L^2(\Omega)}\left\|A^{\frac{3}{2}}\varphi^\perp\right\|_{L^2(\Omega)}$$
$$\leqslant C_w\left\|\Delta_x\varphi^\perp\right\|_{L^2(\Omega)}\left\|A^{\frac{3}{2}}\varphi^\perp\right\|_{L^2(\Omega)}.$$

The latter inequality leads to

$$\left\|A\varphi^{\perp}\right\|_{L^2(\Omega)} \leq C_w \left\|A^{\frac{3}{2}}\varphi^{\perp}\right\|_{L^2(\Omega)}.$$

Substituting that relation into the previous inequality with $\Delta_x u = \Delta_x \varphi^{\perp}$, $\nabla_x u = \nabla_x \varphi^{\perp}$ and using the boundedness $g(s) \leq g^*$, $|g'(s)| \leq L_g$ by Condition 4.1 (i), (iii) and $N_2(t) \leq \|\overline{N}\|_{L^\infty(\Omega)}$, we obtain

$$\frac{d}{dt}\|\Delta_x \varphi^{\perp}(t)\|^2_{L^2(\Omega)} + \frac{d_1}{C_W{}^2}\|\Delta_x \varphi^{\perp}(t)\|^2_{L^2(\Omega)}$$

$$\leq \frac{2d_2{}^2}{d_1}L_g^2\|\overline{N}\|^2_{L^\infty(\Omega)}\|\nabla_x \varphi^{\perp}(t)\|^2_{L^2(\Omega)} + \frac{2d_2{}^2}{d_1}g^{*2}\|\nabla_x N_2(t)\|^2_{L^2(\Omega)}.$$

In the following we assume

$$\sup_{t>0} \|\nabla_x N_2(t)\|_{L^2(\Omega)} \leq C_N < \infty \tag{4.52}$$

and show that under this assumption we have $\sup_{t \geq \tilde{\delta}} \|\Delta_x \varphi^{\perp}(t)\|_{L^2(\Omega)} < \infty$.

Remark 4.12   It remains to show that this assumption holds true in general.

From (4.46) we know that $\sup_{t \geq \delta} \|\nabla_x \varphi^{\perp}(t)\|_{L^2(\Omega)} \leq C_{12}$ and due to Condition 4.3 even $\sup_{t>0} \|\nabla_x \varphi^{\perp}(t)\|_{L^2(\Omega)} \leq C_{13}$ (c.f. Remark 4.9). In summary we have

$$\frac{d}{d}y(t) + \gamma y(t) \leq C_{16} =: \frac{2d_2{}^2}{d_1}L_g^2\|\overline{N}\|^2_{L^\infty(\Omega)}C_{13}^2 + \frac{2d_2{}^2}{d_1}g^{*2}C_N^2$$

where $y(t) := \|\Delta_x \varphi^{\perp}(t)\|^2_{L^2(\Omega)}$ and $\gamma := \frac{d_1}{C_W{}^2}$. Proposition 4.2 yields by integration $\int_{\underline{t}}^t ds$ with $\underline{t} > 0$

$$\|\Delta_x \varphi^{\perp}(t)\|^2_{L^2(\Omega)} \leq \|\Delta_x \varphi^{\perp}(\underline{t})\|^2_{L^2(\Omega)} + \frac{C_{16}}{\gamma} \quad \text{for all } t \geq \underline{t}. \tag{4.53}$$

Theorem 4.1 states that the solution u fulfills $\sqrt{t}\,\Delta_x u \in L^2(0, T; L^2(\Omega))$, i.e., the solution is immediately smoothened. By definition it holds for $\Delta_x u = \Delta_x \varphi^{\perp}$

$$\int_0^T t\|\Delta_x \varphi^{\perp}(t)\|^2_{L^2(\Omega)} \, dt < \infty,$$

which yields the existence of a time $\underline{t} \in (0, T]$ and a constant $C_{17} < \infty$ such that

$$\underline{t}\|\Delta_x \varphi^{\perp}(\underline{t})\|^2_{L^2(\Omega)} \leq C_{17} \Leftrightarrow \|\Delta_x \varphi^{\perp}(\underline{t}))\|_{L^2(\Omega)} \leq \frac{1}{\sqrt{\underline{t}}}C_{17}.$$

This expression is bounded if $\underline{t}$ is bounded away from zero, i.e., if $\underline{t} \geqslant \tilde{\delta} > 0$ for some positive constant $\tilde{\delta} > 0$. Then (4.53) implies

$$\|\Delta_x \varphi^\perp(t)\|^2_{L^2(\Omega)} \leqslant \frac{1}{\tilde{\delta}} C_{17} + \frac{C_{16}}{\gamma} =: C_{18}^2 < \infty \quad \text{for all } t \geqslant \underline{t} \geqslant \tilde{\delta}$$

and consequently

$$\sup_{t \geqslant \tilde{\delta}} \|\Delta_x \varphi^\perp(t)\|_{L^2(\Omega)} \leqslant C_{18}.$$

In order to show the uniform convergence of $u(t)$ to u^∞, at this point it only remains to show that assumption (4.52) is always satisfied. For that the higher regularity of the initial data under the terms of Condition 4.3 is crucial.

First we need to show that the same holds for $\nabla_x N_1(t)$, i.e., $\sup_{t>0} \|\nabla_x N_1(t)\|_{L^2(\Omega)} < \infty$. Application of the gradient to the model Eq. (4.2) leads to

$$\partial_t \nabla_x N_1 = -f'(u)\nabla_x u \, N_1 - f(u)\nabla_x N_1$$

and by multiplying with $\nabla_x N_1$ and integrating over $\int_\Omega dx$ we obtain

$$\frac{1}{2}\frac{d}{dt}\|\nabla_x N_1(t)\|^2_{L^2(\Omega)}$$

$$= -\int_\Omega f'(u(x,t))\nabla_x u(x,t) N_1(x,t)\nabla_x N_1(x,t)\, dx$$

$$- \int_\Omega f(u(x,t))|\nabla_x N_1(x,t)|^2\, dx$$

$$\leqslant \int_\Omega |f'(u(x,t))||\nabla_x u(x,t)||N_1(x,t)||\nabla_x N_1(x,t)|\, dx$$

$$- \int_\Omega f(u(x,t))|\nabla_x N_1(x,t)|^2\, dx. \tag{4.54}$$

Remark 4.13 Recall the decomposition

$$u(x,t) = a_1(t)C_\varphi + \varphi^\perp(x,t).$$

From previous observations we know $u^\infty = a_1^\infty C_\varphi$ and the function $a_1(t)$ in nondecreasing, i.e.

$$a_1(t)C_\varphi \leqslant u^\infty \quad \text{for all } t \geqslant 0. \tag{4.55}$$

Now Condition 4.2 (ii) comes into play, which holds for $u \in [C^-, C^- + \delta]$. According to that we define the following subsets of Ω for all $t \geqslant 0$:

$$\Omega_1(t) := \{x \in \Omega : u(x, t) < C^-\}$$

$$\Omega_2(t) := \{x \in \Omega : u(x, t) \in [C^-, C^- + \delta]\}$$

$$\Omega_3(t) := \{x \in \Omega : u(x, t) > C^- + \delta\}$$

Then for every point of time we have a disjoint partition of Ω:

$$\Omega = \Omega_1(t) \mathbin{\dot{\cup}} \Omega_2(t) \mathbin{\dot{\cup}} \Omega_3(t) \quad \forall t \geqslant 0.$$

Our aim now is to estimate the term $|f'(u(x, t))|$ appearing in (4.54) for the regions $\Omega_1(t)$ and $\Omega_2(t)$. From Condition 4.2 (i) we know that $f(u(x, t))$ is constant zero on $\Omega_1(t)$ and thus it holds

$$|f'(u(x, t))| = 0 \quad \text{on } \Omega_1(t).$$

To obtain an estimate in $\Omega_2(t)$, at this stage we have to distinguish between the two cases

$$\text{I) } u^\infty \leqslant C^- \quad \text{and} \quad \text{II) } u^\infty > C^-.$$

Case I

We first take a look at the case $u^\infty \leqslant C^-$. On $\Omega_2(t)$ we have by (4.55)

$$0 \leqslant m_1 \left(u(x, t) - C^-\right) \leqslant f'\left(u(x, t)\right) \leqslant m_2 \left(u(x, t) - C^-\right)$$

$$\leqslant m_2 \left(u^\infty + \varphi^\perp(x, t) - C^-\right) \leqslant m_2 \varphi^\perp(x, t)$$

$$\Rightarrow |f'(u(x, t))| \leqslant m_2 |\varphi^\perp(x, t)| \quad \text{on } \Omega_2(t).$$

Substituting these estimates into (4.54), we get

$$\frac{1}{2} \frac{d}{dt} \|\nabla_x N_1(t)\|^2_{L^2(\Omega)} \leqslant m_2 \|\overline{N}\|_{L^\infty(\Omega)} \int_{\Omega_2(t)} |\varphi^\perp(x, t)| |\nabla_x u(x, t)| |\nabla_x N_1(x, t)| \, dx$$

$$+ \|\overline{N}\|_{L^\infty(\Omega)} \int_{\Omega_3(t)} |f'(u(x, t))| |\nabla_x u(x, t)| |\nabla_x N_1(x, t)| \, dx$$

$$- \int_\Omega f(u(x, t)) |\nabla_x N_1(x, t)|^2 \, dx. \tag{4.56}$$

The first integral can be further estimated by use of Hölder's inequality:

$$\int_{\Omega_2(t)} |\varphi^\perp(x,t)| |\nabla_x u(x,t)| |\nabla_x N_1(x,t)| \, dx \leq \|\varphi^\perp(t) \nabla_x u(t) \nabla_x N_1(t)\|_{L^1(\Omega)}$$

$$\leq \|\varphi^\perp(t) \nabla_x u(t)\|_{L^2(\Omega)} \|\nabla_x N_1(t)\|_{L^2(\Omega)}$$

$$\leq \|\varphi^\perp(t)\|_{L^4(\Omega)} \|\nabla_x u(t)\|_{L^4(\Omega)} \|\nabla_x N_1(t)\|_{L^2(\Omega)}.$$

From Theorem 4.1 we know that $u(t)$ and $\nabla_x u(t)$ lie in $H^1(\Omega)$ for $t > 0$ and the Sobolev imbedding theorem states $H^1(\Omega) \hookrightarrow L^4(\Omega)$ with constant C_{H^1} since $n \leq 3$. This allows to further estimate

$$\leq C_{H^1}^2 \|\varphi^\perp(t)\|_{H^1(\Omega)} \|\nabla_x u(t)\|_{H^1(\Omega)} \|\nabla_x N_1(t)\|_{L^2(\Omega)}$$

and applying Corollary 1.1 to $\|\varphi^\perp(t)\|_{H^1(\Omega)}$ and $\|\nabla_x u(t)\|_{H^1(\Omega)} = \|\nabla_x \varphi^\perp(t)\|_{H^1(\Omega)}$ leads us to

$$\leq C_{H^1}^2 (C_W^2 + 1) \|\nabla_x \varphi^\perp(t)\|_{L^2(\Omega)} \|\Delta_x \varphi^\perp(t)\|_{L^2(\Omega)} \|\nabla_x N_1(t)\|_{L^2(\Omega)}.$$

For the second integral we artificially insert $\frac{\sqrt{f(u(x,t))}}{\sqrt{f(u(x,t))}}$ which is possible since in $\Omega_3(t)$ it holds $u(x,t) > C^- + \delta$ and with that by Condition 4.2 (iii) we know $f(u(x,t)) > 0$. By Young's inequality we have

$$\int_{\Omega_3(t)} \frac{|f'(u(x,t))|}{\sqrt{f(u(x,t))}} |\nabla_x u(x,t)| \sqrt{f(u(x,t))} |\nabla_x N_1(x,t)| \, dx$$

$$\leq \frac{1}{2} \int_{\Omega_3(t)} \frac{|f'(u(x,t))|^2}{f(u(x,t))} |\nabla_x u(x,t)|^2 \, dx + \frac{1}{2} \int_{\Omega_3(t)} f(u(x,t)) |\nabla_x N_1(x,t)|^2 \, dx$$

and due to Conditions 4.1 (iii) and 4.2 (iii)

$$\leq \frac{1}{2} \frac{L_f^2}{f(C^- + \delta)} \|\nabla_x \varphi^\perp(t)\|_{L^2(\Omega)}^2 + \frac{1}{2} \int_\Omega f(u(x,t)) |\nabla_x N_1(x,t)|^2 \, dx.$$

Substitution of these findings into (4.56) leads to

$$\frac{1}{2} \frac{d}{dt} \|\nabla_x N_1(t)\|_{L^2(\Omega)}^2$$

$$\leq \underbrace{m_2 \|\overline{N}\|_{L^\infty(\Omega)} C_{H^1}^2 (C_W^2 + 1)}_{=:C_{19}} \|\nabla_x \varphi^\perp(t)\|_{L^2(\Omega)} \|\Delta_x \varphi^\perp(t)\|_{L^2(\Omega)} \|\nabla_x N_1(t)\|_{L^2(\Omega)}$$

$$+ \underbrace{\|\overline{N}\|_{L^\infty(\Omega)} \frac{1}{2} \frac{L_f^2}{f(C^- + \delta)}}_{=:\frac{C_{20}}{2}} \|\nabla_x \varphi^\perp(t)\|_{L^2(\Omega)}^2 - \frac{1}{2} \int_\Omega f(u(x,t)) |\nabla_x N_1(x,t)|^2 \, dx.$$

The last term can be omitted and we obtain with Young's inequality

$$\frac{d}{dt}\|\nabla_x N_1(t)\|_{L^2(\Omega)}^2 \leqslant (C_{19} + C_{20})\|\nabla_x \varphi^\perp(t)\|_{L^2(\Omega)}^2$$
$$+ C_{19}\|\Delta_x \varphi^\perp(t)\|_{L^2(\Omega)}^2\|\nabla_x N_1(t)\|_{L^2(\Omega)}^2 .$$

Integration over $\int_0^t ds$ yields with (4.38) under the assumption of Condition 4.3

$$\|\nabla_x N_1(t)\|_{L^2(\Omega)}^2 \leqslant \underbrace{\|\nabla_x N_{1,0}\|_{L^2(\Omega)}^2 + (C_{19} + C_{20})\int_0^\infty \|\nabla_x \varphi^\perp(t)\|_{L^2(\Omega)}^2 dt}_{\leqslant C_{21} < \infty}$$
$$+ C_{19}\int_0^t \|\Delta_x \varphi^\perp(s)\|_{L^2(\Omega)}^2\|\nabla_x N_1(s)\|_{L^2(\Omega)}^2 ds .$$

Now Gronwall's inequality can be applied and we finally obtain with (4.39)

$$\|\nabla_x N_1(t)\|_{L^2(\Omega)}^2 \leqslant C_{21} e^{C_{19}\int_0^\infty \|\Delta_x \varphi^\perp(t)\|_{L^2(\Omega)}^2 dt} < \infty \quad \text{for all } t > 0,$$

i.e., for the case $u^\infty \leqslant C^-$ we achieved

$$\sup_{t > 0} \|\nabla_x N_1(t)\|_{L^2(\Omega)} < \infty . \tag{4.57}$$

This result is now used to show that $\nabla_x N_2(t)$ also stays bounded in $L^2(\Omega)$. For that we apply the gradient to model Eq. (4.3) and get

$$\partial_t \nabla_x N_2 = f'(u)\nabla_x u\, N_1 + f(u)\nabla_x N_1 - g'(u)\nabla_x u\, N_2 - g(u)\nabla_x N_2 .$$

Multiplication by $\nabla_x N_2$ and integration over $\int_\Omega dx$ lead to

$$\frac{1}{2}\frac{d}{dt}\|\nabla_x N_2(t)\|_{L^2(\Omega)}^2$$
$$= \int_\Omega f'(u(x,t))\nabla_x u(x,t)N_1(x,t)\nabla_x N_2(x,t)\, dx$$
$$+ \int_\Omega f(u(x,t))\nabla_x N_1(x,t)\nabla_x N_2(x,t)\, dx$$
$$- \int_\Omega g'(u(x,t))\nabla_x u(x,t)N_2(x,t)\nabla_x N_2(x,t)\, dx$$
$$- \int_\Omega g(u(x,t))|\nabla_x N_2(x,t)|^2\, dx .$$

Condition 4.1 (ii), (iii) yields

$$\frac{1}{2}\frac{d}{dt}\|\nabla_x N_2(t)\|_{L^2(\Omega)}^2 \leq \underbrace{(L_f + L_g)\|\overline{N}\|_{L^\infty(\Omega)}}_{=:C_{22}}\int_\Omega |\nabla_x u(x,t)||\nabla_x N_2(x,t)|\,dx$$

$$+ f^* \int_\Omega |\nabla_x N_1(x,t)||\nabla_x N_2(x,t)|\,dx$$

$$- \int_\Omega g(u(x,t))|\nabla_x N_2(x,t)|^2\,dx\,.$$

Our first aim is to derive the boundedness of $\|\nabla_x N_2(t)\|_{L^2(\Omega)}$ for the fixed interval $t \in (0, t_0]$. Omitting the last integral and using Hölder's together with Young's inequality, we obtain with (4.47) and (4.57)

$$\frac{d}{dt}\|\nabla_x N_2(t)\|_{L^2(\Omega)}^2 \leq C_{22}\|\nabla_x u(t)\|_{L^2(\Omega)}^2 + f^*\|\nabla_x N_1(t)\|_{L^2(\Omega)}^2$$

$$+ (C_{22} + f^*)\|\nabla_x N_2(t)\|_{L^2(\Omega)}^2$$

$$\leq C_{22}\sup_{t>0}\|\nabla_x u(t)\|_{L^2(\Omega)}^2 + f^*\sup_{t>0}\|\nabla_x N_1(t)\|_{L^2(\Omega)}^2$$

$$+ (C_{22} + f^*)\|\nabla_x N_2(t)\|_{L^2(\Omega)}^2$$

$$\leq C + (C_{22} + f^*)\|\nabla_x N_2(t)\|_{L^2(\Omega)}^2 \qquad \text{with } C < \infty\,.$$

Integration over $\int_0^t ds$ with $0 < t \leq t_0$ yields

$$\|\nabla_x N_2(t)\|_{L^2(\Omega)}^2 \leq \|\nabla_x N_{2,0}\|_{L^2(\Omega)}^2 + Ct_0 + (C_{22} + f^*)\int_0^t \|\nabla_x N_2(s)\|_{L^2(\Omega)}^2\,ds$$

and hence with Gronwall's inequality and Condition 4.3

$$\|\nabla_x N_2(t)\|_{L^2(\Omega)}^2 \leq \left(\|\nabla_x N_{2,0}\|_{L^2(\Omega)}^2 + Ct_0\right)e^{(C_{22}+f^*)t_0} < \infty \quad \text{for all } t \in (0, t_0]\,.$$

$$(4.58)$$

In particular this implies

$$\|\nabla_x N_2(t_0)\|_{L^2(\Omega)} < \infty \qquad\qquad (4.59)$$

and we can proceed to the whole time interval $t > 0$.

From Proposition 4.1 we know that for all $x \in \Omega$ it holds $u(x, t) \geqslant \varrho > 0 \ \forall t \geqslant t_0$ and consequently Condition 4.2 (iii), (iv) implies

$$g(u(x, t)) \geqslant g(\overline{\varrho}) > 0 \quad \forall t \geqslant t_0$$

with $\overline{\varrho} = \min(\varrho, \varrho_0)$ as defined earlier.

We use this result and apply Young's inequality to the first two integrals with $\varepsilon_1 = \frac{1}{2C_{22}} g(\overline{\varrho})$ and $\varepsilon_2 = \frac{1}{2f*} g(\overline{\varrho})$. This gives for all $t \geqslant t_0$

$$\frac{1}{2}\frac{d}{dt}\|\nabla_x N_2(t)\|^2_{L^2(\Omega)} + \frac{1}{2} g(\overline{\varrho})\|\nabla_x N_2(t)\|^2_{L^2(\Omega)}$$

$$\leqslant \frac{C^2_{22}}{g(\overline{\varrho})}\|\nabla_x u(t)\|^2_{L^2(\Omega)} + \frac{f^{*2}}{g(\overline{\varrho})}\|\nabla_x N_1(t)\|^2_{L^2(\Omega)}$$

$$\leqslant \frac{C^2_{22}}{g(\overline{\varrho})}\sup_{t>0}\|\nabla_x u(t)\|^2_{L^2(\Omega)} + \frac{f^{*2}}{g(\overline{\varrho})}\sup_{t>0}\|\nabla_x N_1(t)\|^2_{L^2(\Omega)}$$

$$\leqslant C_{23} < \infty \quad \text{due to (4.47) and (4.57)}.$$

Hence Proposition 4.2 can be applied and yields with (4.59)

$$\|\nabla_x N_2(t)\|^2_{L^2(\Omega)} \leqslant \|\nabla_x N_2(t_0)\|^2_{L^2(\Omega)} + \frac{2C_{23}}{g(\varrho)} < \infty \quad \text{for all } t \geqslant t_0,$$

i.e., together with (4.58) we achieved

$$\sup_{t>0}\|\nabla_x N_2(t)\|_{L^2(\Omega)} < \infty.$$

That means for the first case I) $u^\infty \leqslant C^-$ we showed that assumption (4.52) holds for all $t > 0$ and taking it all together we obtain with $C_{24} > 0$ chosen appropriately

$$\sup_{t>0}\|\nabla_x N_2(t)\|_{L^2(\Omega)} < \infty \quad \Rightarrow \quad \sup_{t\geqslant\tilde{\delta}}\|\Delta_x \varphi^\perp(t)\|_{L^2(\Omega)} \leqslant C_{24} < \infty.$$

From the previous observations we know

$$\|u(t) - u^\infty\|_{L^\infty(\Omega)} \leqslant CC_\theta \sup_{t\geqslant\tilde{\delta}}\|\Delta_x \varphi^\perp(t)\|^{1-\theta}_{L^2(\Omega)}\|\nabla_x \varphi^\perp(t)\|^\theta_{L^2(\Omega)} \quad \text{for all } t \geqslant \tilde{\delta}$$

and from the boundedness of $\Delta_x \varphi^\perp(t)$ it follows

$$\|u(t) - u^\infty\|_{L^\infty(\Omega)} \leqslant CC_\theta C^{1-\theta}_{24}\|\nabla_x \varphi^\perp(t)\|^\theta_{L^2(\Omega)} \quad \text{for all } t \geqslant \tilde{\delta}.$$

With (4.37) we finally obtain

$$\|u(t) - u^\infty\|_{L^\infty(\Omega)} \longrightarrow 0 \quad \text{as } t \to \infty$$

and the uniform convergence of $u(t)$ to the constant function u^∞ is shown.

Case II

Now the same considerations are made in the case $u^\infty > C^-$. Again by Condition 4.2 (ii) on $\Omega_2(t)$ it holds

$$m_1(u(x, t) - C^-) \leqslant f'(u(x, t)) \leqslant m_2(u(x, t) - C^-)$$

and this time we take a look at the lower part and integrate over the interval $[C^-, u(x, t)]$:

$$\int_{C^-}^{u(x,t)} m_1(v - C^-)\, dv \leqslant \int_{C^-}^{u(x,t)} f'(v)\, dv = f(u(x, t)) - \underbrace{f(C^-)}_{=0}$$

which leads to

$$f(u(x, t)) \geqslant \frac{1}{2} m_1 \left(u(x, t) - C^- \right)^2 .$$

Due to the condition $u^\infty > C^-$ and the monotonicity of $a_1(t)$, with (4.55) it follows:
For all α with $0 < \alpha < (u^\infty - C^-)$ there exists $T_\alpha > 0$ such that

$$a_1(t)C_\varphi \geqslant C^- + \alpha \quad \text{for all } t \geqslant T_\alpha .$$

Substituting that relation into the previous estimate, we obtain with Young's inequality

$$f(u(x, t)) \geqslant \frac{1}{2} m_1 \left(a_1(t)C_\varphi + \varphi^\perp(x, t) - C^- \right)^2 \geqslant \frac{1}{2} m_1 \left(\alpha + \varphi^\perp(x, t) \right)^2$$

$$\geqslant \frac{1}{2} m_1 \left(\alpha^2 - 2\alpha|\varphi^\perp(x, t)| + |\varphi^\perp(x, t)|^2 \right)$$

$$\geqslant \frac{1}{2} m_1 \left(\frac{1}{2}\alpha^2 - |\varphi^\perp(x, t)|^2 \right) \qquad \text{for all } t \geqslant T_\alpha \text{ and } x \in \Omega_2(t),$$

i.e., this estimates holds for f applied to $u(x, t) \in [C^-, C^- + \delta]$. The boundedness of f from below given by Condition 4.2 (iii) implies

$$f(u(x, t)) \geqslant f(C^- + \delta) \geqslant \frac{1}{2} m_1 \left(\frac{1}{2}\alpha^2 - |\varphi^\perp(x, t)|^2 \right) \qquad \text{for all } t \geqslant T_\alpha \text{ and } x \in \Omega_3(t).$$

We start again with the boundedness of $\nabla_x N_1(t)$ and go back to (4.54). With the preceding observations and $f(u(x,t)) = 0 = f'(u(x,t))$ in $\Omega_1(t)$ we have for $t \geqslant T_\alpha$

$$\frac{1}{2}\frac{d}{dt}\|\nabla_x N_1(t)\|^2_{L^2(\Omega)} \leqslant L_f\|\overline{N}\|_{L^\infty(\Omega)}\int_{\Omega_2(t)\,\cup\,\Omega_3(t)} |\nabla_x(u(x,t))||\nabla_x N_1(x,t)|\,dx$$

$$-\frac{m_1}{4}\alpha^2\int_{\Omega_2(t)\,\cup\,\Omega_3(t)}|\nabla_x N_1(x,t)|^2\,dx + \frac{m_1}{2}\int_{\Omega_2(t)\,\cup\,\Omega_3(t)}|\varphi^\perp(x,t)|^2|\nabla_x N_1(x,t)|\,dx\,.$$

Hölder's and Young's inequality with $\varepsilon = m_1(2L_f\|\overline{N}\|_{L^\infty(\Omega)})^{-1}\alpha^2$ and extension to Ω give

$$\frac{d}{dt}\|\nabla_x N_1(t)\|^2_{L^2(\Omega)} \leqslant \frac{2L_f^2\|\overline{N}\|^2_{L^\infty(\Omega)}}{m_1\alpha^2}\|\nabla_x u(t)\|^2_{L^2(\Omega)}$$

$$+ m_1\|\varphi^\perp(t)\|^2_{L^\infty(\Omega)}\|\nabla_x N_1(t)\|^2_{L^2(\Omega)}\,.$$

For $n \leqslant 3$ there is the Sobolev space imbedding $H^2(\Omega) \hookrightarrow L^\infty(\Omega)$ with constant $C_{H^2} > 0$ and by use of Corollary 1.1 we can further estimate

$$\leqslant \underbrace{\frac{2L_f^2\|\overline{N}\|^2_{L^\infty(\Omega)}}{m_1\alpha^2}}_{=:\,C_{25}}\|\nabla_x\varphi^\perp(t)\|^2_{L^2(\Omega)}$$

$$+ \underbrace{m_1 C_{H^2}^2(C_W^4 + C_W^2 + 1)}_{=:\,C_{26}}\|\Delta_x\varphi^\perp(t)\|^2_{L^2(\Omega)}\|\nabla_x N_1(t)\|^2_{L^2(\Omega)}\,.$$

Integrating over $\int_0^t ds$, we obtain

$$\|\nabla_x N_1(t)\|^2_{L^2(\Omega)} \leqslant \underbrace{\|\nabla_x N_{1,0}\|^2_{L^2(\Omega)} + C_{25}\int_0^\infty \|\nabla_x\varphi^\perp(t)\|^2_{L^2(\Omega)}\,dt}_{\leqslant\,C_{27}\,<\infty}$$

$$+ C_{26}\int_0^t \|\Delta_x\varphi^\perp(s)\|^2_{L^2(\Omega)}\|\nabla_x N_1(s)\|^2_{L^2(\Omega)}\,ds$$

and Gronwall's inequality together with (4.38) leads us to

$$\|\nabla_x N_1(t)\|^2_{L^2(\Omega)} \leqslant C_{27}\,e^{C_{26}\int_0^\infty \|\Delta_x\varphi^\perp(t)\|^2_{L^2(\Omega)}\,dt} < \infty \quad \text{for all } t \geqslant T_\alpha\,,$$

i.e., it holds

$$\sup_{t\,\geqslant\,T_\alpha}\|\nabla_x N_1(t)\|_{L^2(\Omega)} < \infty\,.$$

In analogy to the calculations for $\nabla_x N_2(t)$ in $(0, t_0]$, by use of (4.54), Conditions 4.1 (iii) and 4.3 we also obtain for the bounded interval $(0, T_\alpha]$

$$\sup_{t \in (0,T_\alpha]} \|\nabla_x N_1(t)\|_{L^2(\Omega)} < \infty.$$

All together for case II) $u^\infty > C^-$, we obtain

$$\sup_{t > 0} \|\nabla_x N_1(t)\|_{L^2(\Omega)} < \infty.$$

The boundedness of $\nabla_x N_2(t)$ in $L^2(\Omega)$ is shown in exactly the same way as for case I) $u^\infty \leq C^-$, since for the estimation of $\|\nabla_x N_2(t)\|_{L^2(\Omega)}$ we did not use the relation of u^∞ and C^-. This yields

$$\sup_{t > 0} \|\nabla_x N_2(t)\|_{L^2(\Omega)} < \infty$$

and with that

$$\|u(t) - u^\infty\|_{L^\infty(\Omega)} \longrightarrow 0 \quad \text{as } t \to \infty.$$

In summary for both cases I) and II) we obtained uniform convergence of the solution $u(t)$ to $u^\infty \equiv a_1^\infty C_\varphi$.   □

Convergence Rate

Theorem 4.5 *Under the assumptions of Condition 4.3, the longtime behavior of the ODE solution (N_1, N_2, N_3) can be further characterized:*

Partial swelling: $u^\infty < C^-$
  $\exists\, T_p \geq \tilde{\delta} > 0$ such that for all $x \in \Omega$ it holds

$$N_1(x, t) \equiv N_1(x, T_p) \quad \text{for all } t \geq T_p,$$

and we have the following exponential convergence rates for all $x \in \Omega$ and $t \geq T_p$:

$$N_2(\cdot, t) \xrightarrow{t \to \infty} 0 \qquad\qquad \text{in } \mathcal{O}(e^{-g(\overline{\varrho})t})$$

$$N_3(\cdot, t) \xrightarrow{t \to \infty} \overline{N} - N_1(\cdot, T_p) \qquad \text{in } \mathcal{O}(e^{-g(\overline{\varrho})t}).$$

Complete swelling: $u^\infty > C^-$

$\exists\, T_c \geqslant \tilde{\delta} > 0$ *such that for all* $x \in \Omega$ *and all* $t \geqslant T_c$ *the following exponential convergence rates hold with some constants* $\gamma > 0,\ \zeta > 0,\ \eta > 0$:

$$N_1(\cdot, t) \xrightarrow{t \to \infty} 0 \quad in\ \mathcal{O}\left(e^{-f(C^- + \gamma)t}\right)$$

$$N_2(\cdot, t) \xrightarrow{t \to \infty} 0 \quad in\ \mathcal{O}(e^{-g(\zeta)t})$$

$$N_3(\cdot, t) \xrightarrow{t \to \infty} \overline{N} \quad in\ \mathcal{O}(e^{-\eta t}).$$

Here exponential convergence of $v(t) \to v^\infty$ *is given if there exist some constants* $C > 0$ *and* $k > 0$ *such that*

$$|v(t) - v^\infty| \leqslant Ce^{-kt}.$$

Proof **1.) Partial swelling**

We start with the partial swelling case. By the uniform convergence of $u(x, t)$ to $u^\infty < C^-$ follows the existence of a time $T_p \geqslant \tilde{\delta} > 0$ such that

$$u(x, t) \leqslant C^- \qquad \forall x \in \Omega \quad \forall t \geqslant T_p$$

and consequently

$$f(u(x, t)) = 0 \qquad \forall x \in \Omega \quad \forall t \geqslant T_p. \tag{4.60}$$

From the model Eq. (4.2) it follows immediately for all $x \in \Omega$

$$\partial_t N_1(x, t) = 0 \quad \forall t \geqslant T_p,$$

which implies

$$N_1(x, t) \equiv N_1(x, T_p) \qquad \forall t \geqslant T_p.$$

For Eq. (4.3), by (4.60) and the definition of $\overline{\varrho}$, it holds for all $t \geqslant T_p \geqslant \tilde{\delta}$

$$\partial_t N_2(x, t) = -g(u(x, t))N_2(x, t) \leqslant -g(\overline{\varrho})N_2(x, t),$$

which yields with Gronwall's inequality

$$N_2(x, t) \leqslant N_{2,0}(x)e^{-g(\overline{\varrho})t} \leqslant \|N_{2,0}\|_{L^\infty(\Omega)}e^{-g(\overline{\varrho})t} \quad \forall x \in \Omega \quad \forall t \geqslant T_p \tag{4.61}$$

i.e., we have exponential convergence of $N_2(x, t)$ to 0. From the conservation law (4.8) we know

$$N_3(x, t) = \overline{N}(x) - N_1(x, T_p) - N_2(x, t) \quad \forall t \geqslant T_p$$

and consequently

$$N_{3,p}^{\infty}(x) = \overline{N}(x) - N_1(x, T_p).$$

This gives

$$|N_3(x, t) - N_{3,p}^{\infty}(x)| = N_2(x, t) \quad \text{for } t \geqslant T_p$$

and with the previous result (4.61) it follows

$$|N_3(x, t) - N_{3,p}^{\infty}(x)| \leqslant \|N_{2,0}\|_{L^{\infty}(\Omega)} e^{-g(\overline{\varrho})t} \quad \forall x \in \Omega \quad \forall t \geqslant T_p,$$

i.e., exponential convergence of $N_3(x, t)$ to $N_{3,p}^{\infty}(x)$.

2.) Complete swelling

In the complete swelling case where $u^{\infty} > C^-$, the uniform convergence of $u(x, t)$ to u^{∞} yields the existence of a time $T_c \geqslant \tilde{\delta} > 0$ and a constant $\beta > 0$ such that

$$u(x, t) \geqslant C^- + \beta > C^- \quad \forall x \in \Omega \quad \forall t \geqslant T_c.$$

In order to show the positivity of $f(u(x, t))$, we have to distinguish between two cases in accordance with Condition 4.2:

- $u(x, t) \geqslant C^- + \delta$, where $\delta > 0$ denotes the constant from Condition 4.2, i.e., $\beta \geqslant \delta$. In that case, Condition 4.2 (iii) holds and it follows

$$f(u(x, t)) \geqslant f(C^- + \delta) > 0 \quad \forall x \in \Omega \quad \forall t \geqslant T_c.$$

- $u(x, t) \in (C^-, C^- + \delta)$, which means we are in the case of Condition 4.2 (ii), where we have the relation $f'(u(x, t)) \geqslant m_1\beta > 0$, i.e., the function is strictly increasing and with that

$$f(u(x, t)) \geqslant f(C^- + \beta) > 0 \quad \forall x \in \Omega \quad \forall t \geqslant T_c.$$

In summary we conclude for the situation of complete swelling

$$f(u(x, t)) \geqslant f(C^- + \gamma) > 0 \quad \forall x \in \Omega \quad \forall t \geqslant T_c, \text{ where } \gamma := \min(\beta, \delta)$$

$$(4.62)$$

and in addition by Condition 4.2 (iii), (iv)

$$g(u(x, t)) \geqslant g(\zeta) > 0 \qquad \forall x \in \Omega \quad \forall t \geqslant T_c \,, \text{ where } \zeta := \min(\overline{\varrho}, C^- + \beta) \,.$$
$$(4.63)$$

Substituting relation (4.62) into the model equation

$$\partial_t N_1(x, t) = -f(u(x, t))N_1(x, t) \,,$$

we obtain exponential convergence of $N_1(x, t)$ to 0:

$$N_1(x, t) \leqslant N_{1,0}(x)e^{-f(C^- + \gamma)t} \leqslant \|N_{1,0}\|_{L^\infty(\Omega)}e^{-f(C^- + \gamma)t} \quad \forall x \in \Omega \quad \forall t \geqslant T_c \,.$$

The second ODE

$$\partial_t N_2(x, t) = f(u(x, t))N_1(x, t) - g(u(x, t))N_2(x, t)$$

can be estimated by means of the previous result, Condition 4.1 (ii) and (4.63). This yields for $x \in \Omega$ and $t \geqslant T_c$

$$\partial_t N_2(x, t) \leqslant f^* \|N_{1,0}\|_{L^\infty(\Omega)}e^{-f(C^- + \gamma)t} - g(\zeta)N_2(x, t) \,.$$

Integration over $\int_0^t ds$ gives

$$N_2(x, t) \leqslant \underbrace{\|N_{2,0}\|_{L^\infty(\Omega)} + \frac{f^*}{f(C^- + \gamma)}\|N_{1,0}\|_{L^\infty(\Omega)}}_{=: C_{28} < \infty} - g(\zeta) \int_0^t N_2(x, s) \, ds \,,$$

whence follows with Gronwall's inequality the exponential convergence to 0:

$$N_2(x, t) \leqslant C_{28}e^{-g(\zeta)t} \quad \forall x \in \Omega \quad \forall t \geqslant T_c \,.$$

In analogy to the previous case, the conservation law (4.8) implies

$$N_{3,c}^\infty(x) = \overline{N}(x)$$

and it immediately follows with $\eta := \min\left(f(C^- + \gamma), g(\zeta)\right)$

$$|N_3(x, t) - \overline{N}(x)| \leqslant \left(\|N_{1,0}\|_{L^\infty(\Omega)} + C_{28}\right)e^{-\eta t} \quad \forall x \in \Omega \quad \forall t \geqslant T_c \,.$$

□

Remark 4.14 For the third case $u^\infty = C^-$ no further details about the type of convergence can be made.

Proposition 4.3 *Let the assumptions of Theorem 4.5 hold.*

Then the solution $u(t)$ converges to u^∞ "exponentially fast," i.e., there exists $t_0 > 0$ and constants $k > 0$, $C > 0$ such that

$$\|u(t) - u^\infty\|_{L^2(\Omega)} \leqslant Ce^{-kt_0} \quad \text{for all } t \geqslant t_0,$$

which means we can obtain an arbitrary small norm by the choice of t_0 sufficiently large.

Proof From Theorem 4.5 we know that $N_2(t)$ converges exponentially to 0 in $\mathcal{O}(e^{-kt})$, where $k := \min(g(\overline{\varrho}), g(\zeta))$. In addition, (4.29) states that the main part of the decomposition of $u(x, t)$ is nondecreasing, whence follows

$$\frac{d}{dt} a_1(t) \leqslant d_2 C_\varphi g^* \int_\Omega N_2(x, t)\, dx \leqslant d_2 C_\varphi g^* \lambda(\Omega) \|N_{2,0}\|_{L^\infty(\Omega)} e^{-kt}.$$

Integration over $\int_t^\infty ds$ yields

$$0 \leqslant a_1^\infty - a_1(t) \leqslant \underbrace{d_2 C_\varphi \frac{g^*}{k} \lambda(\Omega) \|N_{2,0}\|_{L^\infty(\Omega)}}_{=:C_{29} < \infty} e^{-kt},$$

i.e.

$$a_1^\infty - a_1(t) = |a_1^\infty - a_1(t)| \leqslant C_{29} e^{-kt} \quad \forall t > 0.$$

By the characterization of the first Fourier coefficient (4.28), the definition (4.26) of C_φ and the fact $u^\infty = a_1^\infty C_\varphi$, it follows for $t > 0$

$$a_1^\infty - a_1(t) = C_\varphi \int_\Omega u^\infty\, dx - C_\varphi \int_\Omega u(x, t)\, dx$$

$$= C_\varphi \left(\|u^\infty\|_{L^1(\Omega)} - \|u(t)\|_{L^1(\Omega)} \right) \leqslant C_{29} e^{-kt}.$$

This relation induces that not $u(t)$ itself, but the mean value over Ω converges exponentially to the mean value of u^∞.

Remark 4.15

1.) Since we do not know the sign of $u^\infty - u(x, t)$ in Ω, we cannot deduce exponential convergence of $u(t)$ to u^∞ in $L^1(\Omega)$.
2.) This kind of exponential mean value convergence is also known from the standard heat equation.

Now we take a look at second part of the decomposition. From (4.35) it follows with Condition 4.1 (ii)

$$\frac{d}{dt}\|\varphi^\perp(t)\|^2_{L^2(\Omega)} + \gamma\|\varphi^\perp(t)\|^2_{L^2(\Omega)} \leqslant C_{30}\|N_2(t)\|^2_{L^2(\Omega)}$$

and by Proposition 4.2 together with the exponential convergence of N_2, we have

$$\|\varphi^\perp(t)\|^2_{L^2(\Omega)} \leqslant \|\varphi^\perp(t_0)\|^2_{L^2(\Omega)}e^{-\gamma(t-t_0)} + C_{30}\int_{t_0}^t \|N_2(s)\|^2_{L^2(\Omega)}\,ds$$

$$\leqslant \|\varphi^\perp(t_0)\|^2_{L^2(\Omega)}e^{-\gamma(t-t_0)} + \frac{C_{30}}{k}\|N_{2,0}\|^2_{L^\infty(\Omega)}\lambda(\Omega)\left(e^{-2kt_0} - e^{-2kt}\right)$$

for every $t \geqslant t_0$. Taking the limit, we obtain

$$\|\varphi^\perp(t)\|^2_{L^2(\Omega)} \leqslant \underbrace{\frac{C_{30}}{k}\|N_{2,0}\|^2_{L^\infty(\Omega)}\lambda(\Omega)\,e^{-2kt_0}}_{=:C_{31}} \quad \text{for every fixed } t_0 > 0 \text{ and } t \geqslant t_0.$$

In summary we obtain with the previous results for $t \geqslant t_0$

$$\|u(t) - u^\infty\|^2_{L^2(\Omega)} = \int_\Omega |a_1(t)C_\varphi + \varphi^\perp(x,t) - a_1^\infty C_\varphi|^2\,dx$$

$$\leqslant 2C_\varphi^2\lambda(\Omega)|a_1(t) - a_1^\infty|^2 + 2\|\varphi^\perp(t)\|^2_{L^2(\Omega)}$$

$$\leqslant 2C_\varphi^2\lambda(\Omega)C_{29}^2e^{-2kt} + 2C_{31}e^{-2kt_0}$$

$$\leqslant (2C_\varphi^2\lambda(\Omega)C_{29}^2 + 2C_{31})e^{-2kt_0}$$

and the proof is complete.

$\square$

4.3 Numerical Analysis: In Vitro

Now we want to verify the obtained mathematical results numerically. As described earlier, the model functions f and g have a sigmoidal shape determined in the following way:

$f(u)$: Transition rate from unswollen to swelling mitochondria dependent on the local Ca^{2+} concentration (Figs. 4.1 and 4.2)

$$f(s) = \begin{cases} 0 & \text{for } s < C^- \\ f^* & \text{for } s > C^+ \\ -\frac{f^*}{2}\cos\left(\frac{s-C^-}{C^+-C^-}\pi\right) + \frac{f^*}{2} & \text{else} \end{cases}$$

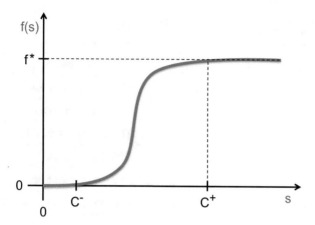

Fig. 4.1 Transition rate f (with permission from AIMS)

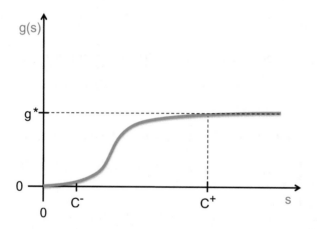

Fig. 4.2 Transition rate g (with permission from AIMS)

$g(u)$: "Dying" term describing the transition of mitochondria in the swelling process to completely swollen ones where the membrane is ruptured and stored Ca^{2+} is released

$$g(s) = \begin{cases} g^* & \text{for } s > C^+ \\ -\frac{g^*}{2} \cos\left(\frac{s}{C^+}\pi\right) + \frac{g^*}{2} & \text{else} \end{cases}$$

Table 4.1 In vitro
simulation: model parameters
I (with permission from
AIMS)

Name	Description	Value
d_1	Diffusion parameter	0.2
d_2	Feedback parameter	30
f^*	Maximal transition rate $N_1 \rightharpoonup N_2$	1
g^*	Maximal transition rate $N_2 \rightharpoonup N_3$	0.1
C^-	Threshold of initiating $N_1 \rightharpoonup N_2$	20
C^+	Saturation threshold	200
$h_x = \frac{1}{N}$	Step size of space discretization	$\frac{1}{40}$
h_t	Step size of time discretization	1

The In Vitro Model

Model Parameters

The model parameters we used for the in vitro simulations are noted in Table 4.1.

Discretization

The domain Ω is discretized with step size $h_x = \frac{1}{N}$ leading to the discrete domain

$$\Omega_h = \{(x_i, x_j)\}_{i,j=0,\dots,N} = \{(ih_x, jh_x) : i = 0, \dots, N, \ j = 0, \dots, N\}$$

of size $(N + 1) \times (N + 1)$.

The time interval will be discretized with time steps of length h_t which results in the discrete time interval

$$T_h = \{t_k\}_{k \in \mathbb{N}_0} = \{kh_t : k \in \mathbb{N}_0\}.$$

On that domain $\Omega_h \times T_h$ we define the grid solution $(u_h, N_{1h}, N_{2h}, N_{3h})$.

Numerical Approximation

The PDE describing the calcium diffusion process is discretized with respect to space by means of the standard finite difference approach. Here the Laplacian is approximated by use of the five point star. Doing so, the PDE is transferred into an ODE and we end up with an ODE system on the discrete domain Ω_h, that shall be solved for the discrete time steps $t_k \in T_h$. Due to the low numerical complexity of the model, this can be easily achieved by using the explicit Euler method. The homogeneous Neumann boundary condition is realized by introducing phantom points in order to calculate the normal derivative at the boundary.

Initial Values

As we pointed out earlier, in the beginning all mitochondria are intact and with that neither in the swelling process nor completely swollen, i.e.

$$N_{1,0} \equiv 1, \quad N_{2,0} \equiv 0, \quad N_{3,0} \equiv 0.$$

For the calcium concentration it is not so clear how to determine the initial state. The initial value u_0 defines the distribution of the added Ca^{2+} amount. At this the rate of diffusion progression as well as the dosage location are of great importance. Therefore one can imagine different possible initial states. Here we take a look at the grid solution u_h and determine initial values u_{0h}. In all cases we assume that the total amount of added Ca^{2+} is the same, i.e.

$$\sum_{x_i, x_j \in \Omega_h} u_{0h}(x_i, x_j) \equiv C_{tot}.$$

1) Highly localized: The total calcium amount C_{tot} is located at one single point (x_k, x_l) in Ω_h:

$$u_{0h}(x_k, x_l) = C_{tot}$$
$$u_{0h}(x_i, x_j) = 0 \qquad \text{for } i \neq k, \, j \neq l, \quad x_i, x_j \in \Omega_h.$$

Figure 4.3 shows two possible distributions.
2) Normally distributed: The initial calcium concentration is determined by a sector of the standard normal distribution. Figure 4.4 depicts the meaning of a sector of

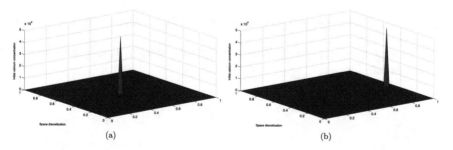

(a)   (b)

Fig. 4.3 Localized initial calcium distribution. (**a**) Center: $(x_k, x_l) = (0.5, 0.5)$. (**b**) Close to boundary: $(x_k, x_l) = (0.5, 0.95)$ (with permission from AIMS)

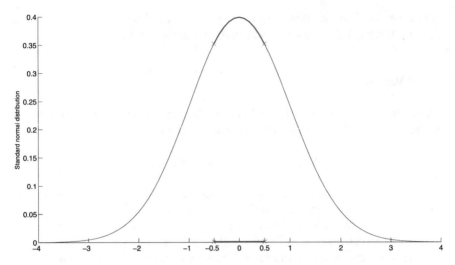

Fig. 4.4 Sector of the 1D standard normal distribution (with permission from AIMS)

the normal distribution for the 1D case. In 2D the density function is given by

$$N(y_1, y_2) = \frac{1}{2\pi} \exp\left(-\frac{1}{2}(y_1^2 + y_2^2)\right)$$

and the sector for the interval $[y_l, y_r]$ is obtained by the translation

$$y \curvearrowright \tilde{y} := (y_r - y_l) \cdot y + y_l \, .$$

Thus the initial Ca^{2+} concentration adapted to the total calcium amount is given by

$$u_{0h}(x_i, x_j) = C_{tot} \cdot \frac{N(\tilde{x}_i, \tilde{x}_j)}{\sum_{x_i, x_j \in \Omega_h} N(\tilde{x}_i, \tilde{x}_j)} \, , \quad x_i, x_j \in \Omega_h \, .$$

Figure 4.5 shows the resulting initial calcium distributions for different sector intervals $[y_l, y_r]$.
3) Constant: As a consequence of complete diffusion the initial calcium distribution is constant, i.e.

$$u_{0h}(x_i, x_j) = \frac{C_{tot}}{(N + 1)^2} \, , \quad x_i, x_j \in \Omega_h \, .$$

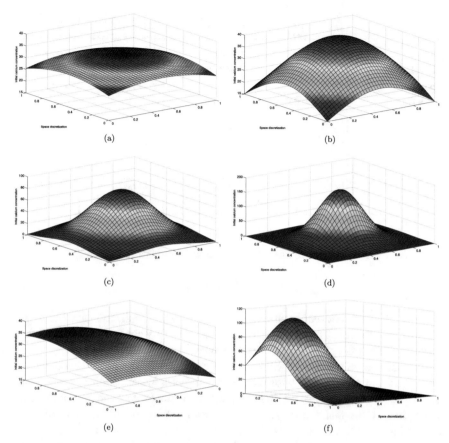

Fig. 4.5 Normally distributed initial calcium distribution (with permission from AIMS). (**a**) $y_l = -0.5$, $y_r = 0.5$. (**b**) $y_l = -1$, $y_r = 1$. (**c**) $y_l = -2$, $y_r = 2$. (**d**) $y_l = -3$, $y_r = 3$. (**e**) $y_l = -0.8$, $y_r = 0.2$. (**f**) $y_l = -1$, $y_r = 3$

Simulation

Now everything is prepared for the numerical simulations, which were done using the commercial software MATLAB.

The following pages show the time evolution of the numerical solutions for five different initial calcium concentrations displayed column by column:

Column 1	Column 2	Column 3	Column 4	Column 5
Localized at point $x = (0.5, 0.5)$	Localized at point $x = (0.5, 0.95)$	Normally distributed on sector $[-1, 3]$	Normally distributed on sector $[-2, 2]$	Normally distributed on sector $[-0.5, 0.5]$
$t = t_1$	$t = t_1$	$t = t_1$	$t = t_1$	$t = t_1$
⋮	⋮	⋮	⋮	⋮
$t = t_{10}$	$t = t_{10}$	$t = t_{10}$	$t = t_{10}$	$t = t_{10}$

For every data collection we used the same total amount of initial Ca^{2+}

$$C_{tot} = 30 \cdot (N + 1)^2,$$

i.e., the difference between each column only lies in the localization and the degree of calcium diffusion. The simulations show that for every constellation this calcium amount is high enough to induce *complete swelling*.

Figures 4.6, 4.7, 4.8, and 4.9 show the evolution of the model variables u_h, N_{1h}, N_{2h}, and N_{3h} column-wise for the five different initial conditions mentioned above. Each row displays the different states at one time step. From the simulations it becomes clear that the choice of the initial calcium distribution is of great importance for the duration as well as the dynamics of the whole process.

One remarkable result is the clearly visible spreading calcium wave. If we compare the dynamics with those of simple diffusion without any feedback, the resulting calcium evolution induced by mitochondrial swelling is indeed completely different.

The numerical simulations also show that at much lower initial calcium concentrations, the model outcome depends on the location and type of the initial calcium distribution. As we can see there, a small change in the initial distribution of Ca^{2+} is enough to shift the behavior from partial to complete swelling (Fig. 4.10).

Comparison with Experimental Data

Up to now, we do not have any experimental data including the spatial development. As we already explained, the existing data measure mitochondrial swelling in terms of light scattering at different calcium concentrations as can again be seen in Fig. 4.11: In order to compare the obtained solutions with the experimental data, we now have to take a look at the combined volume of all three subpopulations. As described in Sect. 3.2, this is obtained by summing the weighted mean values over

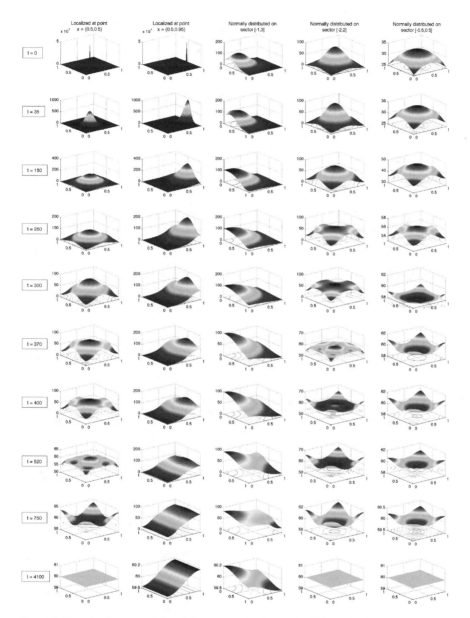

Fig. 4.6 Time development of the **calcium concentration** u_h for different initial calcium concentrations (with permission from AIMS)

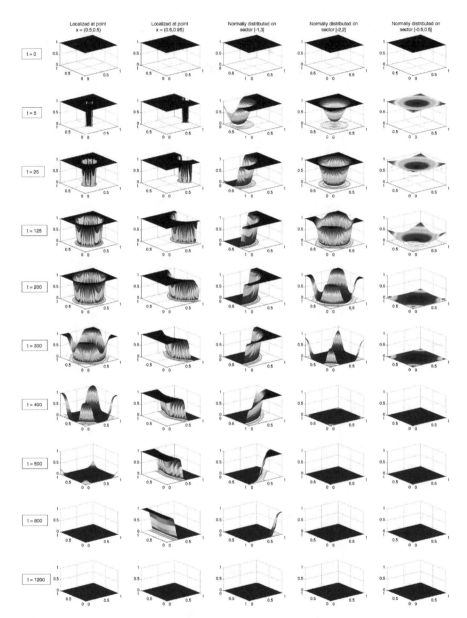

Fig. 4.7 Time development of the **unswollen mitochondrial subpopulation** N_{1h} for different initial calcium concentrations (with permission from AIMS)

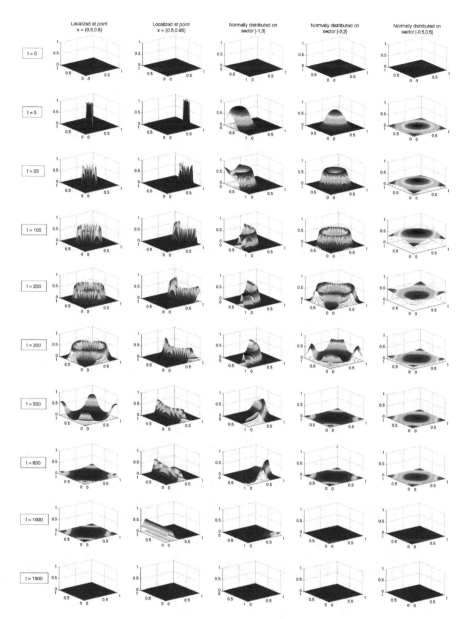

Fig. 4.8 Time development of the **intermediate mitochondrial subpopulation** N_{2h} for different initial calcium concentrations (with permission from AIMS)

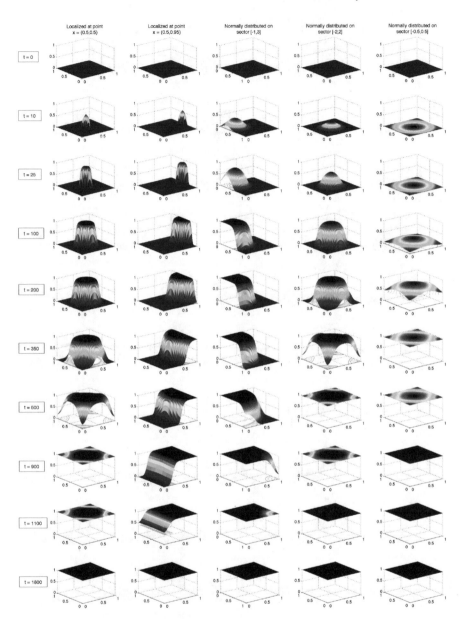

Fig. 4.9 Time development of the **swollen mitochondrial subpopulation** N_{3h} for different initial calcium concentrations (with permission from AIMS)

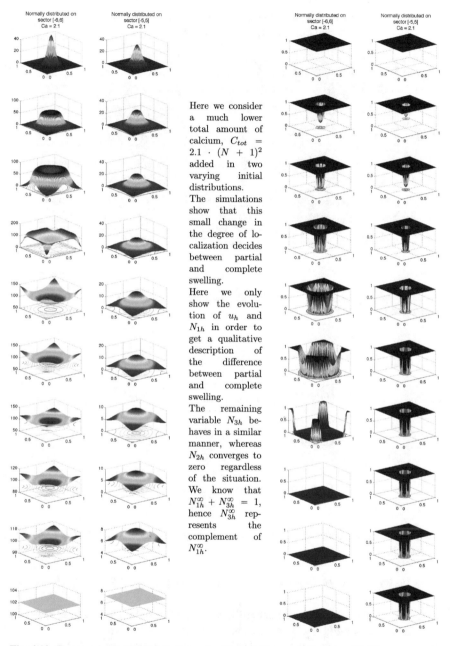

Here we consider a much lower total amount of calcium, $C_{tot} = 2.1 \cdot (N + 1)^2$ added in two varying initial distributions.

The simulations show that this small change in the degree of localization decides between partial and complete swelling.

Here we only show the evolution of u_h and N_{1h} in order to get a qualitative description of the difference between partial and complete swelling.

The remaining variable N_{3h} behaves in a similar manner, whereas N_{2h} converges to zero regardless of the situation. We know that $N_{1h}^{\infty} + N_{3h}^{\infty} = 1$, hence N_{3h}^{∞} represents the complement of N_{1h}^{∞}.

Fig. 4.10   Two varying initial distributions of calcium (with permission from AIMS)

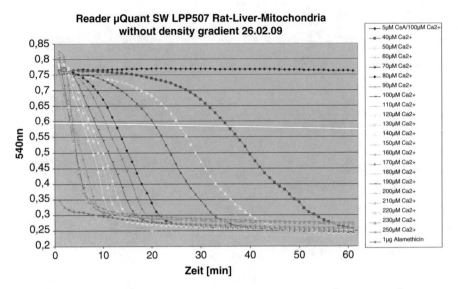

Fig. 4.11 Experimental data of mitochondrial swelling at different Ca^{2+} concentrations (with permission from AIMS)

the whole domain Ω_h. These averages are given by

$$I_{1h}(t_k) = \frac{1}{(N+1)^2} \sum_{x_i, x_j \in \Omega_h} N_{1h}(x_i, x_j, t_k)$$

$$I_{2h}(t_k) = \frac{1}{(N+1)^2} \sum_{x_i, x_j \in \Omega_h} N_{2h}(x_i, x_j, t_k)$$

$$I_{3h}(t_k) = \frac{1}{(N+1)^2} \sum_{x_i, x_j \in \Omega_h} N_{3h}(x_i, x_j, t_k)$$

and with that in accordance with the previously introduced volume Eq. (3.4), the volume is set up as

$$V(t_k) = V_0 I_{1h}(t_k) + k V_p I_{2h}(t_k) + V_p I_{3h}(t_k), \quad t_k \in T_h. \tag{4.64}$$

Here we choose the initial mitochondrial volume V_0 and the volume of completely swollen mitochondria V_p in accordance with [8]. The parameter $0 < k < 1$ determines the volume of mitochondria in the swelling process as a fixed percentage of the final volume V_p. For the simulations we used the values from Table 4.2. As was shown in [6], there is a linear dependence between the total mitochondrial volume and the measured light scattering change. Thus for the simulation, the optical density values are given by $0.95 - V(t_k)$, $t_k \in T_h$ resulting in the biologically reasonable range between 0.25 and 0.75.

Table 4.2 In vitro simulation: model parameters II (with permission from AIMS)

Name	Description	Value
V_0	Initial volume	1.2
V_p	Final volume	1.7
k	Intermediate volume parameter	0.68

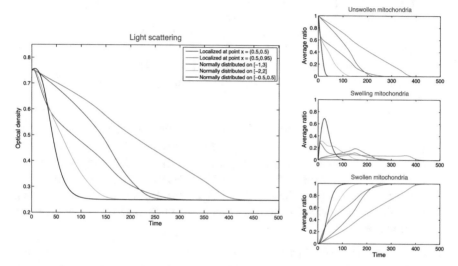

Fig. 4.12 Time development of the average mitochondrial subpopulations and the resulting light scattering values for different initial calcium concentrations (with permission from AIMS)

In the experiment the added Ca^{2+} concentrations are high enough to induce complete swelling. Thus we apply volume formula (4.64) to the solution $(u_h, N_{1h}, N_{2h}, N_{3h})$ of the complete swelling case as depicted in Figs. 4.6, 4.7, 4.8, and 4.9.

Figure 4.12 shows the evolution of each subpopulation in total and displays the resulting optical density values. Here it becomes clear that different initial calcium conditions indeed result in completely different curve shapes. In the following we want to explain the varying curve progressions biologically.

1) *Swelling time*

As one can expect from the 2D simulations, the green line (column $\boxed{2}$) is the slowest one. Due to the most "unfavorable" initial calcium distribution it takes the longest time until all mitochondria are swollen and the lowest optical density value is reached. The shape of the swelling curve displays the linear style of wave propagation.

On the other hand, the black line (column $\boxed{5}$) presents the most "favorable" initial condition and the equilibrium is reached soon. The initial calcium concentration at every point in Ω_h lies above the initiation threshold C^- and thus all mitochondria start swelling immediately. Since there is only little variation over the domain, the whole process is running in a very uniform way. If we take a look at the corresponding "Swelling mitochondria" curve, it is clearly to be seen how the black line increases and decreases very steep within a short period of time. This shows the simultaneity of the incidents because first all mitochondria enter the swelling process at the same time and subsequently complete the swelling process nearly simultaneously. However, the transition $N_2 \rightharpoonup N_3$ is dependent on the local calcium concentration, which is locally not so high compared to the other situations due to the high rate of dispersion. This fact explains why the black swelling curve is the slowest one in the very beginning.

The remaining curves (columns $\boxed{1}$, $\boxed{3}$, and $\boxed{4}$) lie in between and we see that the centered normally distributed initial conditions produce the fastest swelling times.

2) *Number of phases*

Another thing that immediately attracts attention are the different types of curve progression. In particular for the red line (column $\boxed{3}$) the two-phase shape becomes very obvious. This special appearance can be explained by the different dynamics of calcium diffusion. Figure 4.13 studies the two dimensional diffusion process on the quadratic domain Ω. It depicts the dependence of the spreading dynamics on the source location under the assumption of zero flux conditions on the boundary.

In the first image Fig 4.13a the source location is the center of Ω. Since calcium diffusion on a homogenous domain is symmetric, the whole process is radially symmetric and leads to uniform expansion.

The second image Fig. 4.13b considers the situation of the source located close to an edge of the boundary. Here radial symmetry is lost and due to the zero Neumann condition calcium reaching this edge (marked in red) is reflected. Meanwhile calcium is diffusing without interruption in the three remaining directions. The reflection at the boundary has two consequences: (1) Around the reflection area the calcium diffusion is constricted and thus the local concentration here is higher. (2)

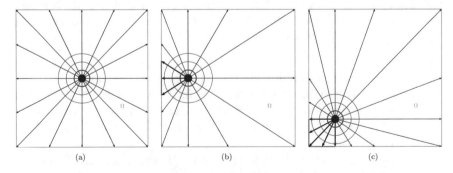

Fig. 4.13 Effects of the source location to the diffusion process in 2D. (**a**) Center. (**b**) Close to edge. (**c**) Close to corner (with permission from AIMS)

This blocked amount of calcium is missing on the other side of the calcium wave and it takes a longer time until all mitochondria are reached. Furthermore this blocked calcium reaches the mitochondria at a later time where they have already been confronted with the freely diffusing calcium. Thus the influence of the "delayed" calcium is diminished.

In the last image Fig. 4.13c the situation is tightened and now the source is located close to a corner, i.e., close to two edges at once. Here the diffusing calcium is reflected and blocked in two directions and can only spread freely in the two remaining directions. Now all the consequences from before apply and are even self-enhanced due to interactions.

Remark 4.16 Of course the same effects occur in the weakened case where the domain is not quadratic but round.

Conclusion

The influence of different initial calcium distributions was experimentally studied in the previously described Fig. 3.4, which is again depicted here. If we compare these data with the simulated swelling curves from Fig. 4.12, we see that we indeed obtain similar curve progressions.

The first case **Volume ratio 1:4** corresponds to the most distributed case of column $\boxed{5}$, whereas the highly localized case **Volume ratio 1:100** shows the same two-phase behavior as the red line (column $\boxed{3}$).

As we saw in the previous simulations, the experimental setting is displayed in the most appropriate way when assuming an initial calcium concentration which is normally distributed on the interval $[-0.5, 0.5]$. Adding different total amounts of calcium in this distribution, we can obtain a qualitative description of the experimental data as it is shown in Fig. 4.14.

In summary, the derived mathematical model gives a realistic description of the mitochondrial swelling taking place in vitro. All experimental data could be verified and furthermore it also yields spatial data. By use of this new model we are now able to study the local processes, which is of major importance in order to understand the underlying biological mechanism in more detail.

4.4 Dirichlet Boundary Conditions

In this section, we continue our analysis of a coupled PDE-ODE model of calcium induced mitochondria swelling. We study the long-term behavior under homogeneous Dirichlet boundary conditions, for which the analytical machinery (as we will see below) that has been developed in Sect. 4.1 does not apply and must be extended. Note that from the biological point of view this kind of boundary

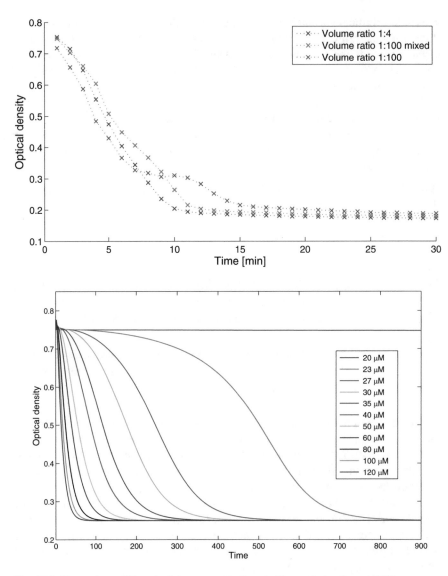

Fig. 4.14 Comparison: different calcium concentrations (with permission from AIMS)

conditions appears if we put some chemical material on the wall that binds calcium ions and hence removes it as a swelling inducer. Hence one can expect that the calcium ion concentration will tend to zero and that in general complete swelling as it was in Sect. 4.1 will not take place as $t \to \infty$. This distinguishes the situation under Dirichlet conditions from the situation under Neumann boundary conditions that were analyzed in Sect. 4.1. Thus, the main goal of this section is to prove rigorously this hypothesis.

As in Sect. 4.1 we analyze the swelling of mitochondria on a bounded domain $\Omega \subset \mathbb{R}^n$ for $n = 2, 3$. The initial calcium concentration $u(x, 0)$ describes the added amount of Ca^{2+} to induce the swelling process. This leads to the following coupled PDE-ODE system determined by the nonnegative model functions f and g:

$$\partial_t u = d_1 \Delta_x u + d_2 g(u) N_2 \tag{4.65}$$

$$\partial_t N_1 = -f(u) N_1 \tag{4.66}$$

$$\partial_t N_2 = f(u) N_1 - g(u) N_2 \tag{4.67}$$

$$\partial_t N_3 = g(u) N_2 \tag{4.68}$$

with diffusion constant $d_1 \geqslant 0$ and feedback parameter $d_2 \geqslant 0$.

The boundary conditions are given by

$$u = 0 \quad \text{on } \partial\Omega. \tag{4.69}$$

The initial conditions are specified as

$$u(x, 0) = u_0(x), \quad N_1(x, 0) = N_{1,0}(x), \quad N_2(x, 0) = N_{2,0}(x), \quad N_3(0.x) = N_{3,0}(x).$$

Note that the total mitochondrial population

$$\bar{N}(x, t) := N_1(x, t) + N_2(x, t) + N_3(x, t)$$

does not change in time, that is, $\partial_t \bar{N}(x, t) = 0$, and is given by the sum of the initial data:

$$\bar{N}(x, t) = \bar{N}(x) := N_{1,0}(x) + N_{2,0}(x) + N_{3,0}(x) \; \forall t \geqslant 0 \; \forall x \in \Omega.$$

The role of model functions f and g as well as calcium evolution $u(x, t)$ is the same as was in detail explained in Sect. 4.1. We now give precise assumptions on f and g.

Condition 4.4 The model functions $f : \mathbb{R} \to \mathbb{R}$ and $g : \mathbb{R} \to \mathbb{R}$ have the following properties:

1. Nonnegativity:

$$f(s) \geqslant 0 \quad \forall s \in \mathbb{R},$$
$$g(s) \geqslant 0 \quad \forall s \in \mathbb{R}.$$

2. Boundedness:

$$f(s) \leqslant f^* < \infty \quad \forall s \in \mathbb{R},$$
$$g(s) \leqslant g^* < \infty \quad \forall s \in \mathbb{R}$$

with $f^*, g^* > 0$.

3. Lipschitz continuity:

$$|f(s_1) - f(s_2)| \leq L_f |s_1 - s_2| \quad \forall s_1, s_2 \in \mathbb{R},$$
$$|g(s_1) - g(s_2)| \leq L_g |s_1 - s_2| \quad \forall s_1, s_2 \in \mathbb{R}$$

with $L_f, L_g \geq 0$.

Condition 4.5 Let the model functions $f : \mathbb{R} \to \mathbb{R}$ and $g : \mathbb{R} \to \mathbb{R}$ fulfil Condition 4.4. In addition we assume that there exist constants $C^- > 0$, $m_1 > 0$, $m_2 > 0$, $\delta_0 > 0$, and $\varrho_0 > 0$ such that the following assertions hold:

1. Starting threshold:

$$f(s) = 0 \quad \forall s \leq C^-$$
$$g(s) = 0 \quad \forall s \leq 0$$

2. Smoothness near starting threshold $[C^-, C^- + \delta]$:

$$|f'(s)| \leq m_f(s - C^-) \quad \forall s \in [C^-, C^- + \delta],$$
$$|g'(s)| \leq m_g s \qquad\qquad \forall s \in [0, \rho_0]$$

3. Lower bounds:

$$f(s) \geq f(C^- + \delta_0) > 0 \quad \forall s \geq C^- + \delta_0,$$
$$g(s) \geq g(\varrho_0) > 0 \qquad\qquad \forall s \geq \varrho_0 > 0.$$

4. Dominance of g: There exists a constant $B > 0$ such that

$$f(s) \leq Bg(s) \quad \forall s \in [0, \infty).$$

Remark 4.17 It is not difficult to see that Condition (4.6) is satisfied for the functions $f, g \in C^2(\mathbb{R}^1_+)$, $f'(C^-) = 0$ and f, g are monotone increasing functions with $f(0) = g(0) = 0$.

Note that the mathematical analysis in Sect. 4.1 is highly dependent on the chosen Neumann boundary conditions. As already mentioned above, the homogeneous Dirichlet problem refers to the scenario where u diffuses across the boundary of the domain into the external space and is immediately removed from there, e.g., by fast dilution and diffusion. This means that one can expect that more and more Ca^{2+} is lost during the process and, as we will show, we end up with $u = 0$, before swelling can be completed. Thus, we study the PDE-ODE system

$$\partial_t u = d_1 \Delta_x u + d_2 g(u) N_2 \tag{4.70a}$$
$$\partial_t N_1 = -f(u) N_1 \tag{4.70b}$$

$$\partial_t N_2 = f(u)N_1 - g(u)N_2 \tag{4.70c}$$

$$\partial_t N_3 = g(u)N_2 \tag{4.70d}$$

with

$$u = 0 \text{ on } \partial\Omega. \tag{4.70e}$$

Our aim now is to study the effect of these boundary conditions.

In the following let $(u^D, N_1^D, N_2^D, N_3^D)$ denote the solution of (4.70a)–(4.70e). It turns out that the change of boundary conditions, which describes how the physical system is connected to the outside world, has a big impact on the mathematical results. There are three main difficulties in the mathematical analysis of the asymptotic behavior of solutions for $t \to \infty$ that prevent us from carrying over the proof for homogeneous Neumann conditions (see Sect. 4.1) in a straightforward manner:

1. Hopf's maximum principle cannot be applied anymore.
2. The corresponding eigenvalue problem has different properties.
3. Wirtinger's inequality does not hold anymore.

For the convenience of the reader, these will be explored in the sequel.

Hopf's Maximum Principle

One main property that was used to study the asymptotic behavior of solutions of the homogeneous Neumann problem was the boundedness of $u(x, t)$ from below by a positive constant $\varrho > 0$ for sufficiently large $t \geq t_0$ (see Proposition 4.1). In order to obtain this lower bound, it was crucial to create a contradiction by use of Hopf's maximum principle, which states that if nonnegative solutions attain 0 at the boundary, then the normal derivative at the boundary cannot be zero. This fact was contrary to the assumption of homogenous Neumann boundary conditions; however, it does not yield a contradiction in the Dirichlet case anymore.

Thus, this argument cannot be used to show that u^D stays away from zero and consequently the transition function $g(u^D)$ is not assured to be positive (which is also obvious from the homogeneous Dirichlet conditions). In order to show the convergence result $N_2^{D,\infty}(x) \equiv 0$ in Sect. 4.1, we needed the Neumann boundary condition, which now does not hold anymore. Due to $g(u^D)$ not necessarily being positive, we cannot deduce that all mitochondria entering the state N_2^D will be transferred to N_3^D.

Corresponding Eigenvalue Problem

In Sect. 4.1 we also showed for the Neumann problem that the calcium ion concentration u strongly converges to a constant C. To prove this fact we essentially used the orthogonal decomposition

$$u(x, t) = a_1(t)\varphi_1(x) + \varphi^{\perp}(x, t)$$

with $\varphi_1(x)$ being the first eigenfunction of the corresponding eigenvalue problem

$$-\Delta_x \varphi_j(x) = \lambda_j \varphi_j(x), \quad x \in \Omega$$
$$\partial_\nu \varphi_j|_{\partial\Omega} = 0.$$

For this type of boundary condition it is known that $\varphi_1(x)$ is constant and with that we could show

$$u(x, t) \to u^{\infty}(x) = \lim_{t\to\infty} a_1(t)\,\varphi_1(x) \equiv C \quad \text{as } t \to \infty.$$

On the contrary for the corresponding Dirichlet eigenvalue problem

$$-\Delta_x \varphi_j^D(x) = \lambda_j^D \varphi_j^D(x), \quad x \in \Omega$$
$$\varphi_j^D|_{\partial\Omega} = 0$$

we can only deduce that the first eigenfunction is positive, but not that it is constant. Even if we could show that the orthogonal complement $\varphi^{D,\perp}$ converges to 0 as in the Neumann case, this would still not allow us to obtain information about the limit function $u^{D,\infty}(x)$. Thus the orthogonal decomposition is only helpful for the Neumann problem.

Wirtinger's Inequality

In order to show some of the convergence results in Sect. 4.1, it was necessary to estimate $\|\varphi^{\perp}(t)\|_{L^2(\Omega)}$ by $\|\nabla_x \varphi^{\perp}(t)\|_{L^2(\Omega)}$. As a result of the orthogonality between $\varphi^{\perp}(t)$ and the constant eigenfunction $\varphi_1(x) \equiv C_\varphi$ in the Neumann case, we could apply Wirtinger's inequality to obtain these estimates.

However, for the Dirichlet problem this is not the case anymore. But this problem can be easily solved, since due to the boundary condition we are now looking at functions from the Sobolev space $H_0^1(\Omega)$. In that space the well-known Poincaré inequality is valid and we can work with the analogous estimate

$$\|v\|_{L^2(\Omega)} \leqslant C_P \|\nabla_x v\|_{L^2(\Omega)} \quad \text{for } v \in H_0^1(\Omega).$$

By use of this inequality we can show that in the Dirichlet case $\|u^D(t)\|_{L^2(\Omega)}$ stays bounded for all $t > 0$. Multiplying (4.70a) by u^D and integrating over $\int_\Omega dx$, we get

$$\frac{1}{2}\frac{d}{dt}\|u^D(t)\|^2_{L^2(\Omega)} + d_1\|\nabla_x u^D(t)\|^2_{L^2(\Omega)} = d_2\int_\Omega g(u^D(x,t))N_2^D(x,t)u^D(x,t)\,dx\,.$$

(4.71)

Remark 4.18 Integration by parts yields

$$\int_\Omega \Delta_x u\,u\,dx = \int_{\partial\Omega}\nabla_x u\,u\,\vec{n}\,dS - \int_\Omega \nabla_x u\,\nabla_x u\,dx = -\|\nabla_x u\|^2_{L^2(\Omega)},\quad \text{for}\quad u \in H^2(\Omega)$$

regardless of the choice of homogenous boundary condition.

We emphasize that equality (4.71) is exactly the same as in the Neumann case with $\varphi^\perp$ replaced by u^D. From here we can proceed in the same way by use of Poincaré's instead of Wirtinger's inequality and obtain

$$\frac{d}{dt}\|u^D(t)\|^2_{L^2(\Omega)} + \underbrace{\frac{d_1}{C_P^2}}_{=:\gamma > 0}\|u^D(t)\|^2_{L^2(\Omega)} \leqslant \frac{d_2^2 C_P^2}{d_1}\|g(u^D(t))N_2^D(t)\|^2_{L^2(\Omega)}\quad (4.72)$$

$$\leqslant \frac{d_2^2 C_P^2}{d_1}\lambda|\Omega|g^{*2}\|\bar{N}\|^2_{L^\infty(\Omega)} =: C < \infty\,.$$

It follows from (i) of Proposition 4.2 with $y(t) = \|U^D(t)\|^2_{L^2(\Omega)}$, $t_0 = 0$, $t_1 = \infty$, $\gamma_0 = \gamma$, and $a(t) = C$ together with (4.72) that

$$\|u^D(t)\|^2_{L^2(\Omega)} \leqslant \|u_0\|^2_{L^2(\Omega)} + \frac{C}{\gamma} =: C_D < \infty\quad \text{for all } t > 0,$$

hence

$$\sup_{t>0}\|u^D(t)\|_{L^2(\Omega)} \leqslant C_D\,.$$

From (4.72) we cannot directly deduce $u^D(x,t) \to 0$ in $L^2(\Omega)$ as we did in the analogous Neumann case for $\varphi^\perp$. For that we need $\|N_2^D\|_{L^1(0,\infty;L^1(\Omega))} < \infty$ in order to have $\int_0^\infty a(t)\,dt < \infty$ with $a(t) := \|g(u^D(t))N_2^D(t)\|^2_{L^2(\Omega)}$. This was shown in Sect. 4.1 and is, amongst others, based essentially on the Neumann boundary condition, which is not valid anymore.

However, we can derive the L^2-summability of $g(u^D(t))N_2^D(t)$ in a different way:

Since the nonnegativity of N_1, N_2, N_3, and u also holds true in the Dirichlet case, it follows from (4.70d) that $N_3^D(x, t)$ is nondecreasing in t for each $x \in \Omega$ and is bounded due to the conservation law: $N_3^D(x, t) \leq \|\bar{N}\|_{L^\infty(\Omega)}$ (see Sect. 4.1). Thus the sequence converges and we have $N_3^D(x, t) \to N_3^{D,\infty}(x) \leq \|\bar{N}\|_{L^\infty(\Omega)}$ as $t \to \infty$. Hence integrating (4.70d) over $\int_0^t dt$ and passing to the limit, we obtain

$$\int_0^\infty g(u^D(x, t))N_2^D(x, t)\, dt = N_3^{D,\infty}(x) - N_{3,0}(x) \leq \|\bar{N}\|_{L^\infty(\Omega)} < \infty \quad \forall x \in \Omega.$$

This yields $g(u^D)N_2^D \in L^1(0, \infty; L^1(\Omega))$, since we can estimate

$$\int_0^\infty \int_\Omega g(u^D(x, t))N_2^D(x, t)\, dx\, dt \leq \|\bar{N}\|_{L^\infty(\Omega)} < \infty.$$

Finally we arrive at the desired result

$$\int_0^\infty a(t)\, dt = \int_0^\infty \|g(u^D(t))N_2^D(t)\|_{L^2(\Omega)}^2\, dt$$

$$\leq g^* \|\bar{N}\|_{L^\infty(\Omega)} \|g(u^D)N_2^D\|_{L^1(0,\infty; L^1(\Omega))} < \infty.$$

Hence (ii) of Proposition 4.2 with $y(t) = \|u^D(t)\|_{L^2(\Omega)}$, $t_0 = 0$, $t_1 = \infty$, $\gamma_0 = \gamma$, and

$$a(t) = \frac{d_2^2 C_\rho^2}{d_1^2} \|g(u^D(t))N_2^D(t)\|_{L^2(\Omega)}^2$$

assures

$$u^D(x, t) \xrightarrow{t \to \infty} u^{D,\infty}(x) \equiv 0 \quad \text{strongly in } L^2(\Omega).$$

Convention 4.1 *For notational convenience, we further write N_1, N_2, N_3, and u instead of N_1^D, N_2^D, N_3^D, and u^D.*

The mitochondrial swelling process depends on the local calcium ion concentration. If the initial concentration u_0 stays below the initiation threshold C^- at all points $x \in \Omega$, and if $N_2(x, 0) \equiv 0$, then no swelling will happen and we have $N_i(x, t) \equiv N_{i,0}(x) \; \forall x \in \Omega, i = 1, 2, 3$.

As we already mentioned in Sect. 4.5 another possible scenario that can be observed in experiments is "partial swelling." This occurs when the initial calcium concentration lies above C^- in a small region, but decreases due to diffusion and falls below this threshold eventually. This leads $N_1(x, t) = N_1(x, T_1) \; \forall t \geq T_1$, N_2 decreases, and N_3 can increase to a positive value.

Thirdly, and this is the case for homogeneous Neumann conditions, cf Sect. 4.1, if the initial calcium distribution together with the influence of the positive feedback is sufficiently high, then "complete swelling" occurs which means $N_1(x, t) \to 0$ and $N_3(x, t) \to \bar{N}(x)$ for all $x \in \Omega$.

As it was shown in Sect. 4.1, in the case of homogeneous Neumann boundary conditions it holds that $N_2(x, t) \to 0$. In order to prove $u(x, t) \to 0$ as $t \to +\infty$ in $L^\infty(\Omega)$ which leads to partial swelling of the mitochondria, we need to obtain

$$\sup_{t>0} \|\nabla_x N_1\|_{L^2(\Omega)} \leqslant C_* \quad \text{and} \quad \sup_{t>0} \|\nabla_x N_2\|_{L^2(\Omega)} \leqslant C_*$$

for some C_*. To this end, let us recall that, in the same manner as for the Neumann boundary conditions, one can easily obtain

1. For any $u_0 \in L^2(\Omega)$, $N_{i,0} \in L^\infty(\Omega)$, $i = 1, 2, 3$, there exists a unique global solution (u, N_1, N_2, N_3) satisfying

$$u \in C([0, T], L^2(\Omega)),$$

$$\sqrt{t}\partial_t u \in C([0, T], L^2(\Omega)),$$

$$\sqrt{t}\Delta_x u \in L^2([0, T], L^2(\Omega)),$$

$$N_i \in L^\infty([0, T], L^\infty(\Omega)), \ i = 1, 2, 3$$

   for all $T > 0$.
2. nonnegativity of solutions
3. Furthermore, the following properties hold

$$\|u(t)\|_{L^2(\Omega)} + \|\nabla_x u(t)\|_{L^2(\Omega)} \to 0 \quad \text{as} \quad t \to \infty, \tag{4.73a}$$

$$\int_0^\infty \|\nabla_x u\|_{L^2(\Omega)}^2 \leqslant C < +\infty, \tag{4.73b}$$

$$\sup_{t>\delta} \|\nabla_x u\|_{L^2(\Omega)} \leqslant C_\delta < +\infty, \tag{4.73c}$$

$$\int_\delta^\infty \|\Delta_x u\|_{L^2(\Omega)}^2 \leqslant C_\delta \tag{4.73d}$$

   for any $\delta > 0$.

As we mentioned above, in order to prove partial swelling, we need to obtain first of all $u(x, t) \to 0$ as in $L^\infty(\Omega)$. In order to prove this we need the following condition.

Condition 4.6 Let the assumption of Condition 4.5 holds. In addition, we assume more regularity of the initial data

$$N_{1,0} \in H^1(\Omega), \ N_{2,0} \in H^1(\Omega).$$

The main theorem of this section is the following:

Theorem 4.6 *Let the assumptions of Condition 4.6 hold. Then it holds that*

$$\sup_{t>0} \|\nabla_x N_1(t)\|_{L^2(\Omega)} < \infty,$$

$$\sup_{t>0} \|\nabla_x N_2(t)\|_{L^2(\Omega)} < \infty, \tag{4.74}$$

$$\sup_{t \geqslant \delta} \|\Delta_x u(t)\|_{L^2(\Omega)} \leqslant C_\delta < \infty, \tag{4.75}$$

where θ is a positive number, and Ω is a bounded domain in $\mathbb{R}^n$, $n \leqslant 3$. Furthermore we have uniform convergence.

$$\|u(t)\|_{L^\infty(\Omega)} \to 0 \; as \; t \to \infty. \tag{4.76}$$

Proof First, we show that (4.75) follows from (4.74). Indeed, let $A = -\Delta_x$ be the Laplacian with $D(A) = H^2(\Omega) \cap H_0^1(\Omega)$. It is well known that the operator A is a positive and self-adjoint. Thus we can define its square root $A^{\frac{1}{2}}$ which inherits the property of self-adjointness. It is easy to see that

$$\|A^{\frac{1}{2}} v\|_{L^2(\Omega)}^2 = (A^{\frac{1}{2}} v, A^{\frac{1}{2}} v) = \|\nabla_x v\|_{L^2(\Omega)}^2$$

and PDE for the calcium ions can be written in terms of A:

$$\partial_t u + d_1 A u = d_2 g(u) N_2$$

and the application of $A^{\frac{1}{2}}$ to this equation gives

$$\partial_t A^{\frac{1}{2}} u + d_1 A^{\frac{3}{2}} u = d_2 A^{\frac{1}{2}} (g(u) N_2).$$

Multiplying by $A^{\frac{3}{2}} u$ and integrating over Ω we obtain due to $A^{\frac{1}{2}}$ being self-adjoint

$$\frac{1}{2} \frac{d}{dt} \|Au\|_{L^2(\Omega)}^2 + d_1 \|A^{\frac{3}{2}} u\|_{L^2(\Omega)}^2 = d_2 (A^{\frac{1}{2}} (g(u) N_2), A^{\frac{3}{2}} u)_{L^2(\Omega)}.$$

Hence, using the Cauchy-Schwarz and Young's inequality we have

$$\frac{1}{2} \frac{d}{dt} \|Au\|_{L^2(\Omega)}^2 + \frac{d_1}{2} \|A^{\frac{3}{2}} u\|_{L^2(\Omega)}^2 \leqslant \frac{d_2^2}{2d_1} \|\nabla_x (g(u(t)) N_2(t))\|_{L^2(\Omega)}^2 \tag{4.77}$$

$$\leqslant \frac{d_2^2}{d_1} \|g'(u(t)) \nabla_x u(t) N_2(t)\|_{L^2(\Omega)}^2 + \frac{d_2^2}{d_1} \|g(u(t)) \nabla_x N_2(t)\|_{L^2(\Omega)}^2.$$

Due to the Poincare inequality

$$\| - \Delta_x u\|^2_{L^2(\Omega)} = (-\Delta_x u, -\Delta_x u)_{L^2(\Omega)} = (Au, Au)_{L^2(\Omega)}$$

$$= (A^{\frac{1}{2}} u, A^{\frac{3}{2}} u) \leqslant \|A^{\frac{1}{2}} u\|_{L^2(\Omega)} \|A^{\frac{3}{2}} u\|_{L^2(\Omega)}$$

$$\leqslant \|\nabla_x u\|_{L^2(\Omega)} \|A^{\frac{3}{2}} u\|_{L^2(\Omega)}.$$

On the other hand, for $u|_{\partial \Omega} = 0$ we have

$$\|\nabla_x u\|^2_{L^2(\Omega)} = (-\Delta_x u, u)_{L^2(\Omega)} \leqslant \|\Delta_x u\|_{L^2(\Omega)} \|u\|_{L^2(\Omega)} \leqslant C_w \|\Delta_x u\|_{L^2(\Omega)} \|\nabla_x u\|_{L^2(\Omega)}.$$

Hence for $u|_{\partial \Omega} = 0$ we have $\|\nabla_x u\|_{L^2(\Omega)} \leqslant C_w \|\Delta_x u\|_{L^2(\Omega)}$. Substituting the last inequality into the above ones we have

$$\|\Delta_x u\|_{L^2(\Omega)} \leqslant C_w \|A^{\frac{3}{2}} u\|_{L^2(\Omega)}.$$

Therefore from (4.77) it follows that

$$\frac{1}{2} \frac{d}{dt} \| - \Delta_x u\|^2_{L^2(\Omega)} + \frac{d_1}{2} \|\Delta_x u\|^2_{L^2(\Omega)} \leqslant \frac{d_2^2}{d_1} \|g'(u(t))\nabla_x u(t) N_2(t)\|^2_{L^2(\Omega)}$$

$$+ \frac{d_2^2}{d_1} \|g(u(t))\nabla_x N_2(t)\|^2_{L^2(\Omega)}.$$

Using the hypothesis and the boundedness $g(s) \leqslant g^*$, $|g'(s)| \leqslant L_g$, and $N_2(x, t) \leqslant \|\bar{N}\|_{L^\infty(\Omega)}$, we obtain

$$\frac{d}{dt} \|\Delta_x u\|^2_{L^2(\Omega)} + C_{1*} \|\Delta_x u\|^2_{L^2(\Omega)} \leqslant C_{2*} \|\nabla_x u\|^2_{L^2(\Omega)} + C_{3*} \|\nabla_x N_2\|^2_{L^2(\Omega)},$$

where C_{i*}, $i = 1, 2, 3$ are some constants.

By virtue of (4.73a), if $\sup_{t>0} \|\nabla_x N_2(t)\|_{L^2(\Omega)} < C$, then (i) of Proposition 4.2 with $y(t) = \|\Delta_x u(t)\|^2_{L^2(\Omega)}$, $t_0 = \delta$, $t_1 = \infty$, $\gamma_0 = C_{1*}$ and $a(t) = C_{2*} \|\nabla_x u(t)\|^2_{L^2(\Omega)} + C_{3*} \|\nabla_x N_2(t)\|^2_{L^2(\Omega)}$ assures that

$$\sup_{t \geqslant \delta} \|\Delta_x u(t)\|_{L^2(\Omega)} \leqslant C\delta.$$

Next we aim at proving $\sup_{t>0} \|\nabla_x N_2(t)\|_{L^2(\Omega)} < \infty$. To this end, we first apply the ∇_x to the equation for $N_1(x, t)$:

$$\partial_t \nabla_x N_1 = -f'(u)\nabla_x u N_2 - f(u)\nabla_x N_1$$

and by multiplying with $\nabla_x N_1$ and integrating over Ω we obtain

$$\frac{1}{2}\frac{d}{dt}\|\nabla_x N_1(t)\|_{L^2(\Omega)}^2 = -\int_\Omega f'(u(x,t))\nabla_x u(x,t) \cdot \nabla_x N_1(x,t)N_1(x,t)dx$$

(4.78)

$$-\int_\Omega f(u(x,t))|\nabla_x N_1(x,t)|^2 dx.$$

We define the following subsets of Ω for all $t \geq 0$:

$$\Omega_1(t) := \{x \in \Omega|\, u(x,t) < C^-\},$$
$$\Omega_2(t) := \{x \in \Omega|\, u(x,t) \in [C^-, C^- + \delta_0]\},$$
$$\Omega_3(t) := \{x \in \Omega|\, u(x,t) > C^- + \delta_0\}.$$

Then using $|f'(u)| \leq m_f|u - c^-| \leq m_f|u|$ for $x \in \Omega_2(t)$ we have

$$\int_\Omega f'(u)\nabla_x N_1 \cdot \nabla_x u N_1 dx$$

$$= \int_{\Omega_2(t)} f'(u)\nabla_x N_1 \cdot \nabla_x u N_1 dx + \int_{\Omega_3(t)} f'(u)\nabla_x N_1 \cdot \nabla_x u N_1 dx$$

$$\leq \|N_1\|_{L^\infty(\Omega)}m_f \int_{\Omega_2(t)} |u||\nabla_x u||\nabla_x N_1|dx + \|N_1\|_{L^\infty(\Omega)}$$

$$\int_{\Omega_3(t)} \frac{|f'(u)|}{\sqrt{f(u)}}|\nabla_x u|\sqrt{f(u)}|\nabla_x N_1|dx$$

$$\leq \|N_1\|_{L^\infty(\Omega)}m_f \int_\Omega |u||\nabla_x u||\nabla_x N_1|dx + \frac{L_f^2}{f(C^- + \delta_0)}$$

$$\int_\Omega |\nabla_x u|^2 dx + \frac{1}{4}\int_\Omega f(u)|\nabla_x N_1|^2 dx.$$

Next we apply the gradient to model equation for $N_2(x,t)$:

$$\partial_t \nabla_x N_2 = f'(u)\nabla_x u N_1 + f(u)\nabla_x N_1 - g'(u)N_2\nabla_x u N_2 - g(u)\nabla_x N_2$$

Multiplication by $\nabla_x N_2$ and integration over Ω lead to

$$\frac{1}{2}\frac{d}{dt}\|\nabla_x N_2\|^2_{L^2(\Omega)} = \int_\Omega f'(u(x,t))\nabla_x u \, N_1 \cdot \nabla_x N_2 dx + \int_\Omega f(u)\nabla_x N_1 \cdot \nabla_x N_2 dx$$

$$(4.79)$$

$$-\int_\Omega g'(u(x,t))\nabla_x u N_2(x,t) \cdot \nabla_x N_2(x,t) dx$$

$$-\int_\Omega g(u(x,t))|\nabla_x N_2|^2 dx.$$

We define

$$J_1 := \int_\Omega |f'(u)||\nabla_x u||N_1||\nabla_x N_2|dx, \quad J_2 := \int_\Omega |f(u(x,t))||\nabla_x N_1||\nabla_x N_2|dx,$$

$$J_3 := \int_\Omega |g'(u(x,t))|\nabla_x u \nabla_x N_2 N_2(x,t)dx, \quad J_4 := \int_\Omega g(u)|\nabla_x N_2(x,t)|^2 dx.$$

It is obvious that for arbitrary $A > 0$ (which will be fixed later)

$$J_2 \leqslant \frac{A}{4}\int_\Omega f(u)|\nabla_x N_1|^2 dx + \frac{1}{A}\int_\Omega f(u)|\nabla_x N_2|^2 dx.$$

To estimate J_1 and J_3 we make the same trick as above namely:

$$|J_1| \leqslant \int_{\Omega_2} |f'(u)||\nabla_x u||\nabla_x N_2||N_2|dx + \int_{\Omega_3} |f'(u)||\nabla_x u||\nabla_x N_2||N_2||\nabla_x u|dx$$

$$\leqslant \|N_2\|_{L^\infty(\Omega)} m_f \int_\Omega |u||\nabla_x N_2||\nabla_x u|dx + \|N_2\|_{L^\infty(\Omega)}$$

$$\int_{\Omega_3} \frac{f'(u)}{\sqrt{f(u)}} \cdot \sqrt{f(u)} \cdot |\nabla_x N_2|dx$$

$$\leqslant \|N_2\|_{L^\infty(\Omega)} m_f \int_\Omega |u||\nabla_x N_2||\nabla_x u|dx + \frac{L_f^2}{f(C^- + \delta_0)}$$

$$\int_\Omega |\nabla_x u|^2 dx + \frac{1}{4}\int_\Omega f(u)|\nabla_x N_2|^2 dx.$$

$$|J_3| \leqslant \int_{\Omega_2} |g'(u(x,t))||\nabla_x u||\nabla_x N_2||N_2|dx$$

$$\leqslant \|N_2\|_{L^\infty(\Omega)} \int_{\{x\in\Omega|u(x,t)\leqslant\rho_0\}} |g'(u(x,t))||\nabla_x u||\nabla_x N_2|dx$$

$$+ \|N_2\|_{L^\infty(\Omega)} \int_{\{x\in\Omega|u(x,t)\geqslant\rho_0\}} |g'(u(x,t))||\nabla_x u||\nabla_x N_2|dx$$

$$\leqslant \|N_2\|_{L^\infty(\Omega)} \int_{\{x\in\Omega|\ |u|\leqslant\rho_0\}} m_g|u||\nabla_x u||\nabla_x N_2|dx$$

$$+ \|N_2\|_{L^\infty(\Omega)} \int_{\{x\in\Omega|u(x,t)\geqslant\rho_0\}} \frac{g'(u)}{\sqrt{g(u)}}|\nabla_x u|\cdot\sqrt{g(u)}\cdot|\nabla_x N_2|dx$$

$$\leqslant m_g\|N_2\|_{L^\infty(\Omega)} \int_\Omega |u||\nabla_x u||\nabla_x N_2|dx + \frac{L_g^2}{g(\rho_0)}$$

$$\int_\Omega |\nabla_x u|^2 dx + \frac{1}{4}\int_\Omega g(u)|\nabla_x N_2|^2 dx,$$

where C_* and $\tilde{C}_*$ are some positive constants. Hence we have

$$\partial_t\|\nabla_x N_1(t)\|_{L^2(\Omega)}^2 \leqslant \tilde{C}_1 \int_\Omega |u||\nabla_x u||\nabla_x N_1|dx + \tilde{C}_1\int_\Omega |\nabla_x u|^2 dx$$

$$- \int_\Omega \frac{3}{4}f(u)|\nabla_x N_1|^2 dx,$$

$$\partial_t\|\nabla_x N_2(t)\|_{L^2(\Omega)}^2 \leqslant \tilde{C}_2 \int_\Omega |u||\nabla_x u||\nabla_x N_2|dx + \tilde{C}_2\int_\Omega |\nabla_x u|^2 dx$$

$$+ \int_\Omega \frac{A}{4}f(u)|\nabla_x N_1|^2 dx$$

$$+ \frac{1}{A}\int_\Omega f(u)|\nabla_x N_2|^2 dx - \frac{3}{4}\int_\Omega g(u)|\nabla_x N_2|^2 dx,$$

where $\tilde{C}_1$ and $\tilde{C}_2$ are some positive constants.

Let $y(t) := A\|\nabla_x N_1(t)\|_{L^2(\Omega)}^2 + \|\nabla_x N_2(t)\|_{L^2(\Omega)}^2$ and $\tilde{C}_3 = \max(\tilde{C}_1, \tilde{C}_2)$. Then we have

$$\partial_t[A\|\nabla_x N_1(x,t)\|_{L^2(\Omega)}^2 + \|\nabla_x N_2(x,t)\|_{L^2(\Omega)}^2]$$

$$\leqslant \tilde{C}_3 \int_\Omega |u||\nabla_x u|(A|\nabla_x N_1| + |\nabla_x N_2|)dx + \tilde{C}_3(A+1)\int_\Omega |\nabla_x u|^2 dx$$

$$- \frac{3A}{4}\int_\Omega f(u)|\nabla_x N_1|^2 dx + \frac{A}{4}\int_\Omega f(u)|\nabla_x N_1|^2 dx$$

$$+ \frac{1}{A}\int_\Omega f(u)|\nabla_x N_2|^2 dx - \frac{3}{4}\int_\Omega g(u)|\nabla_x N_2|^2 dx.$$

Choosing $A = 2B$ (B is given by Condition 4.6), we obtain from the first inequality that

$$\partial_t[A\|\nabla_x N_1\|^2_{L^2(\Omega)} + \|\nabla_x N_2\|^2_{L^2(\Omega)}]$$

$$\leqslant \tilde{C}_3 \int_\Omega |u||\nabla_x u|(A|\nabla_x N_1| + |\nabla_x N_2|)dx + \tilde{C}_3(A+1)\int_\Omega |\nabla_x u|^2 dx$$

$$\leqslant 2\tilde{C}_3\|u\|_{L^4}\|\nabla_x u\|_{L^4}(A\|\nabla_x N_1\|^2_{L^2(\Omega)} + \|\nabla_x N_2\|^2_{L^2(\Omega)})^{\frac{1}{2}} + \tilde{C}_3(A+1)\|\nabla_x u\|^2_{L^2(\Omega)}$$

$$\leqslant 2\tilde{C}_3\|\nabla_x u\|_{L^2(\Omega)}\|\Delta_x u\|_{L^2(\Omega)}(A\|\nabla_x N_1\|^2_{L^2(\Omega)} + \|\nabla_x N_2\|^2_{L^2(\Omega)})^{\frac{1}{2}}$$

$$+ \tilde{C}_3(A+1)\|\nabla_x u\|^2_{L^2(\Omega)}$$

$$\leqslant \tilde{C}_3\|\nabla_x u\|^2_{L^2(\Omega)}(A\|\nabla_x N_1\|^2_{L^2(\Omega)} + \|\nabla_x N_2\|^2_{L^2(\Omega)}) + \tilde{C}_3\|\Delta_x u\|^2_{L^2(\Omega)}$$

$$+ \tilde{C}_3(A+1)\|\nabla_x u\|^2_{L^2(\Omega)}.$$

Since $y(t) = A\|\nabla_x N_1(t)\|^2_{L^2(\Omega)} + \|\nabla_x N_2(t)\|^2_{L^2(\Omega)}$, one can see that $y(t)$ satisfies

$$y'(t) \leqslant \tilde{C}_3\|\nabla_x u\|^2_{L^2(\Omega)}y(t) + \tilde{C}_3\|\Delta_x u\|^2_{L^2(\Omega)} + \tilde{C}_3(A+1)\|\nabla_x u\|^2_{L^2(\Omega)}.$$

Integrating the last inequality over (δ, t) we obtain

$$y(t) \leqslant y(\delta) + \int_\delta^t \tilde{C}_3\|\nabla_x u\|^2_{L^2(\Omega)}y(s)ds$$

$$+ \int_\delta^t (\tilde{C}_3\|\Delta_x u\|^2_{L^2(\Omega)} + \tilde{C}_3(A+1)\|\nabla_x u\|^2_{L^2(\Omega)})ds.$$

Hence, the Gronwall inequality leads to

$$y(t) \leqslant [y(\delta) + \tilde{C}_3\int_\delta^\infty \|\Delta_x u\|^2_{L^2(\Omega)}ds$$

$$+ \tilde{C}_3(A+1)\int_0^\infty \|\nabla_x u\|^2_{L^2(\Omega)}ds]e^{\int_0^\infty \tilde{C}_3\|\nabla_x u\|^2_{L^2(\Omega)}ds}.$$

Here in view of (4.78) and (4.79), we get

$$\frac{d}{dt}\|\nabla_x N_1(t)\|_{L^2(\Omega)} \leqslant L_f\|\bar{N}\|_{L^\infty}\|\nabla_x u\|_{L^2(\Omega)},$$

$$\frac{d}{dt}\|\nabla_x N_2(t)\|_{L^2(\Omega)} \leqslant (L_f + L_g)\|\bar{N}\|_{L^\infty}\|\nabla_x u\|_{L^2(\Omega)} + f^*\|\nabla_x N_1(t)\|_{L^2(\Omega)}.$$

Integrating these inequalities over $(0, \delta)$ with respect to t and using the fact that $N_{1,0}, N_{2,0} \in H^1(\Omega)$ and (4.73b), we can deduce the a priori bound for $y(\delta)$. Therefore due to the estimates (4.73b), (4.73d) we have

$$\sup_{t>0} \|\nabla_x N_2(t)\|_{L^2(\Omega)} \leqslant C < \infty.$$

Now we are in a position to prove (4.76). Indeed, due to the interpolation theorem [10], there exists $\theta \in (0, 1)$ such that we have $\forall t \geqslant \tilde{\delta}$

$$\|u(t)\|_{L^\infty(\Omega)} \leqslant C \|u(t)\|_{C^\alpha(\Omega)} \leqslant C C_\theta \|\nabla_x u(t)\|_{L^2(\Omega)}^\theta \|\Delta_x u(t)\|_{L^2(\Omega)}^{1-\theta}.$$

From (4.73a) we know that $\|\nabla_x u(t)\|_{L^2(\Omega)} \to 0$ as $t \to \infty$ and $\sup_{t \geqslant \delta} \|\Delta_x u(t)\|_{L_2(\Omega)} \leqslant C_\delta$. Hence

$$u(t) \to 0 \text{ as } t \to \infty \text{ in } L^\infty(\Omega).$$

This proves Theorem 4.6.   □

Corollary 4.2 *Let the assumptions of Theorem 4.6 hold. Then we have partial swelling.*

Indeed, by the uniform convergence of $u(x, t)$ to $u^\infty \equiv 0 < C^-$ follows that the existence of $T_p > 0$ such that

$$u(x, t) \leqslant C^- \ \forall x \in \Omega, \ t \geqslant T_p$$

and consequently $f(u(x, t)) \equiv 0 \ \forall x \in \Omega \ \forall t \geqslant T_p$. Then from the model equation for $N_1(x, t)$ it follows immediately for all $x \in \Omega$

$$\partial_t N_1(x, t) = 0 \ \forall t \geqslant T_p,$$

which implies that

$$N_1(\cdot, t) \equiv N_1(\cdot, T_p)$$

However, in contrast to Neumann BC we cannot prove that $N_2(x, t) \to 0$. We aim to show that $\exists \rho > 0$ such that $N_2^\infty(x) \geqslant \rho > 0$ a.e. $x \in \Omega$, where N_2^∞ denotes the limit of N_2 as $t \to \infty$.

Theorem 4.7 *Let the assumptions that $g(s)$ is monotone increasing in $[0, \rho_0]$, where $\rho_0 > 0$ is the same as in Condition 4.5, be satisfied. Assume that there exist $T_1 \in [0, \infty)$ and $\rho_1 > 0$ such that $N_2(x, T_1) \geqslant \rho_1$ for a.e. $x \in \Omega$. Then there exists some $\rho > 0$ such that $N_2^\infty(x) \geqslant \rho > 0$ for a.e. $x \in \Omega$.*

Proof Let $\mathbb{B}_R = \{x \in \mathbb{R}^N \mid |x| \leqslant R\}$ be such that $\Omega \Subset \mathbb{B}_R$, and let $\varphi_1^R(x)$ be the first eigenvalue of

$$\begin{cases} -\Delta_x\varphi = \lambda_1^R\varphi(x) \text{ in } x \in \mathbb{B}_R, \\ \varphi|_{\partial\mathbb{B}_R} = 0 \end{cases}$$

such that $\|\varphi_1^R\|_{L^\infty(\Omega)} = 1$. It is well known that $\varphi_1^R(x) > 0$, and there exists C_R such that $\varphi_1^R|_\Omega \geqslant \frac{C_R}{\rho_0} > 0$, where ρ_0 is the same as in Condition 4.5. Since $\|u(t)\|_{L^\infty(\Omega)} \to 0$ as $t \to \infty$, hence there exists $T_0 > 0$ such that, for all $t \geqslant T_0$, we have

$$\begin{cases} d_2 m_g \|N_2\|_{L^\infty(\Omega)} \|u(t)\|_{L^\infty(\Omega)} \leqslant d_1\lambda_1^R, \\ \|u(t)\|_{L^\infty(\Omega)} \leqslant \min(C_R, \rho_0, C^-). \end{cases}$$

Then for $\forall t \geqslant T_0$ we have

$$|d_2 g(u(x,t))N_2(x,t)| \leqslant d_2\frac{1}{2}m_g|u| \cdot |u|\|N_2(\cdot,t)\|_{L^\infty(\Omega)} \leqslant \frac{1}{2}d_1\lambda_1^R|u| \qquad (4.80)$$

due to the inequality $|g'(u(x,t))| \leqslant m_g|u|$ for all $|u| \leqslant \rho_0$. Let $\overline{u}(x,t) := \lambda(t)\varphi_1^R(x)$, where $\lambda(t)$ is a solution of

$$\begin{cases} \lambda'(t) + \frac{1}{2}\lambda(t)d_1\lambda_1^R = 0, \quad \text{for } t \geqslant T_0, \\ \lambda(T_0) = \rho_0. \end{cases} \qquad (4.81)$$

From (4.81) it follows that

$$\lambda(t) = \rho_0 e^{-\frac{1}{2}d_1\lambda_1^R(t-T_0)} \text{ for } t \geqslant T_0$$

Then

$$|\overline{u}(x,t)| = |\lambda(t)\varphi_1^R(x)| \leqslant \rho_0 \text{ for } t \geqslant T_0$$

and $\overline{u}(x,t)$ satisfies

$$\begin{cases} \partial_t\overline{u} = d_1\Delta_x\overline{u} + \frac{1}{2}d_1\lambda_1^R\overline{u}, \\ \overline{u}|_{t=t_0} = \rho_0, \ \overline{u}|_{\partial\Omega} = 0 \end{cases}$$

and consequently $\overline{u}(x,t)$ is a supersolution of (4.70a) due to the comparison theorem. Indeed, $u(x,T_0) \leqslant C_R = \frac{C_R}{\rho_0} \cdot \rho_0 \leqslant \varphi_1^R(x) \cdot \lambda(T_0) = \overline{u}(x,T_0)$ and $u|_{\partial\Omega} = \overline{u}|_{\partial\Omega} = 0$. Moreover, due to the inequality $|\overline{u}(x,t)| \leqslant \rho_0$ for $t \geqslant T_0$ and

(4.80) we have

$$\partial_t \overline{u} - d_1 \Delta_x \overline{u} - d_2 g(\overline{u}) N_2(x,t) \geqslant \partial_t \overline{u} - \Delta_x \overline{u} - \frac{1}{2} d_1 \lambda_1^R \overline{u}(x,t) = 0.$$

Hence $\partial_t \overline{u} - d_1 \Delta_x \overline{u} - d_2 g(\overline{u}) N_2(x,t) \geqslant \partial_t u - d_1 \Delta_x u - d_2 g(u) N_2(x,t)$. Thus by the comparison theorem, we have

$$u(x,t) \leqslant \overline{u}(x,t) \quad \text{for a.e. } x \in \Omega.$$

After these preliminaries we are in a position to prove the assertion. Indeed, since

$$u(x,t) \leqslant \overline{u}(x,t) \leqslant \lambda(t) \leqslant \rho_0 e^{-\frac{1}{2} d_1 \lambda_1^R (t-T_0)} \quad \forall t \geqslant T_0$$

and $g(s)$ is monotone increasing functions in $s \in [0, \rho_0]$ we obtain

$$g(u(x,t)) \leqslant g\left(\rho_0 e^{-\frac{1}{2} d_1 \lambda_1^R (t-T_0)}\right) \leqslant \frac{1}{2} m_g \rho_0^2 e^{-d_1 \lambda_1^R (t-T_0)}.$$

Moreover, taking into account that $\|u(x,t)\|_{L^\infty(\Omega)} \leqslant C^-$ for $t \geqslant T_0$, we have $f(u(x,t)) = 0$ and consequently

$$\partial_t N_2(x,t) = -g(u(x,t)) N_2(x,t).$$

Thus, $\partial_t N_2(x,t) \geqslant -\frac{1}{2} m_g \rho_0^2 e^{-d_1 \lambda_1^R (t-T_0)} N_2(x,t)$ for all $t \geqslant T_0$. From the last inequality it follows that

$$\int_{T_0}^t \frac{dN_2}{N_2} \geqslant -\int_{T_0}^t \frac{1}{2} m_g \rho_0^2 e^{-d_1 \lambda_1^R (t-T_0)} dt.$$

Hence

$$\log \frac{N_2(x,t)}{N_2(x,T_0)} \geqslant \left[\frac{1}{2} \frac{1}{d_1 \lambda_1^R} m_g \rho_0^2 e^{-d_1 \lambda_1^R (s-T_0)} \right]_{s=T_0}^{s=t} = \frac{m_g \rho_0^2}{2 d_1 \lambda_1^R} [e^{-d_1 \lambda_1^R (t-T_0)} - 1].$$

Since $\partial_t N_2(x,t) < 0$ for $t \geqslant T_0$ we have

$$\frac{N_2(x,t)}{N_2(x,T_0)} < 1 \quad \text{and} \quad \log \frac{N_2(x,t)}{N_2(x,T_0)} < 0.$$

Consequently,

$$\frac{N_2(x,T_0)}{N_2(x,t)} \leqslant e^{\frac{m_g \rho_0^2}{2 d_1 \lambda_1^R} [1 - e^{-d_1 \lambda_1^R (t-T_0)}]}$$

and as a result of

$$\frac{N_2(x, T_0)}{N_2^\infty(x)} \leqslant e^{\frac{m_g \rho_0^2}{2d_1 \lambda_1^R}} \quad \text{as } t \to \infty.$$

Hence,

$$N_2^\infty(x) \geqslant N_2(x, T_0) e^{-\frac{m_g \rho_0^2}{2d_1 \lambda_1^R}}.$$

Then to complete the proof, it suffices to show that

$$\inf_{x \in \Omega} N_2(T_0, x) > 0.$$

For the case where $T_0 \leqslant T_1$, it is clear that we can repeat the same argument above with T_0 replaced by T_1. Hence the conclusion is obvious.

As for the case where $T_1 < T_0$, we note that N_2 satisfies

$$\partial_t N_2(x, t) \geqslant -g^* N_2(x, t) \quad \text{for all } t > 0,$$

which implies

$$\partial_t \left[N_2(x, t) e^{g^*(t-T_1)} \right] \geqslant 0 \quad \Rightarrow \quad N_2(x, t) e^{g^*(t-T_1)} \geqslant N_2(x, T_1).$$

Thus we obtain

$$N_2(x, T_0) \geqslant e^{-g^*(T_0-T_1)} N_2(x, T_1) \geqslant e^{-g^*(T_0-T_1)} \rho_1 > 0 \quad \text{for a.e. } x \in \Omega.$$

$\square$

Remark 4.19 The following two assumptions are sufficient conditions for $N_2(x, T_1) \geqslant \rho_1 > 0$ for a.e. $x \in \Omega$ with $T_1 \in [0, \infty)$.

(1) $\inf_{x \in \Omega} N_2(x, 0) > 0$.
(2) $\inf_{x \in \Omega} u(x, 0) > C^-$, $\inf_{x \in \Omega} N_1(x, 0) > 0$ and $f(s)$ is strictly monotone increasing on $[C^-, C^- + \delta_0]$.

In fact, it is clear that we can take $T_1 = 0$ for the case (1). As for the case (2), by assumption, there exist $t_1 > 0$ and $\bar{\rho}_1 > 0$ such that

$$f(u(x, t)) N_1(x, t) \geqslant \bar{\rho}_1 > 0 \quad \text{for all } t \in [0, t_1].$$

Here we note that N_2 satisfies

$$\partial_t N_2(x, t) \geq f(u(x, t)) N_1(x, t) - g^* N_2(x, t).$$

Hence we get

$$N_2(x, t) \geq e^{-g^* t} \int_0^t e^{g^* s} f(u(x, s)) N_1(x, s) ds \geq \bar{\rho}_1 t \quad \text{for all } t \in (0, t_1].$$

Thus we can take $T_1 = t_1$ and $\rho_1 = \bar{\rho}_1 t_1$.

4.5 Numerical Simulation

For numerical simulations we have to specify functions $f(u)$ and $g(u)$ that satisfy Conditions 4.4 and 4.5. Following Sect. 4.1 we choose

$$f(u) = \begin{cases} 0, & 0 \leq u \leq C^-, \\ \frac{f^*}{2}\left(1 - \cos\frac{(u-C^-)\pi}{C^+-C^-}\right), & C^- \leq u \leq C^+, \\ f^*, & u > C^+, \end{cases} \tag{4.82}$$

and

$$g(u) = \begin{cases} \frac{g^*}{2}\left(1 - \cos\frac{u\pi}{C^+}\right), & 0 \leq u \leq C^+, \\ g^*, & u > C^+. \end{cases} \tag{4.83}$$

As domain $\Omega \subset \mathbb{R}^2$ we choose a disc with diameter 1. In our simulations we use the initial data

$$u(x, 0) = 5C^+\left(1 - \sqrt{x_1^2 + x_2^2}^6\right), \quad x \in \Omega$$

and

$$N_1(x, 0) = 1, \quad N_2(x, 0) = 0, \quad N_3(x, 0) = 0, \quad x \in \Omega,$$

i.e., we assume that initially swelling has not yet been initiated.

We conduct five simulations of the model using the parameters in Table 4.3 and varying the diffusion coefficient d_1 from 0.04 to 0.2, where the maximum value is the one given in Sect. 4.1. In all cases the simulation is terminated if steady state is

Table 4.3 Default parameter values, from Sect. 4.1 (with permission from AIMS)

Parameter	Symbol	Value	Remark
Lower (initiation) swelling threshold	C^-	20	Varied in some simulations
Upper (maximum) swelling threshold	C^+	200	
Maximum transition rate for $N_1 \rightarrow N_2$	f^*	1	
Maximum transition rate for $N_2 \rightarrow N_3$	g^*	0.1	
Diffusion coefficient	d_1	0.2	Varied in some simulations
Feedback parameter	d_2	30	

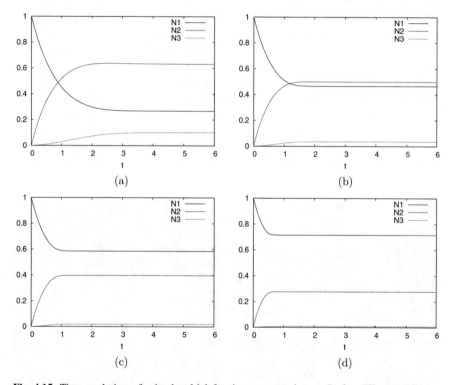

Fig. 4.15 Time evolution of mitochondrial fractions, averaged over Ω, for different diffusion coefficients d_1. (**a**) $d_1 = 0.04$. (**b**) $d_1 = 0.08$. (**c**) $d_1 = 0.12$. (**d**) $d_1 = 0.2$ (with permission from AIMS)

reached, i.e., when

$$\|u^k - u^{k+1}\|_\infty < 10^{-3} \Delta_x t_k \left(10^{-10} + \|u^k\|_\infty\right),$$

$$\|N_j^k - N_j^{k+1}\|_\infty < 10^{-3} \Delta_x t_k \left(10^{-10} + \|N_j^k\|_\infty\right), \quad j = 1, 2, 3,$$

where the index k denotes the kth time-step.

In Figure 4.15 we plot for four different values of the diffusion coefficient d_1 the time evolution of the mitochondrial fractions, averaged over Ω, e.g.

$$N_{j,avg}(t) = \frac{\int_\Omega N_j(x, t)dx}{\int_\Omega dx}, \quad j = 1, 2, 3.$$

In all simulations, $N_{1,avg}(t)$ is monotonically decreasing as per (4.66) and $N_{3,avg}(t)$ is increasing as per (4.68). Higher diffusion coefficients means faster removal of u from the system. Accordingly, the smaller the diffusion coefficient, the more swelling takes place. This manifests itself in lower average values for N_1 and higher average values for N_3. The dependence of $N_{2,avg}$ on d_1 is not as straightforward to predict, since (4.67) has both a source and a sink term. Our simulations show $N_{2,avg}(t)$ is monotonically increasing and that higher diffusion coefficients d_1 lead to lower $N_{2,avg}(t)$.

In Fig. 4.16 we visualize the solution upon termination for $d_1 = 0.16$, attained at $t = 10.6$. As predicted by the analysis above, u is eventually depleted, in our case dropping by 17 orders of magnitude, to reach $u < 3 \cdot 10^{-14} \approx 0$ everywhere. The density of intact, unswollen mitochondria is highest at the boundary, where u is always below C^-, and remains at $N_1 = 1$ in a thin layer there. It decreases toward the center of the domain, attaining $N_1 \approx 0.48$ there as its lowest value. For N_2

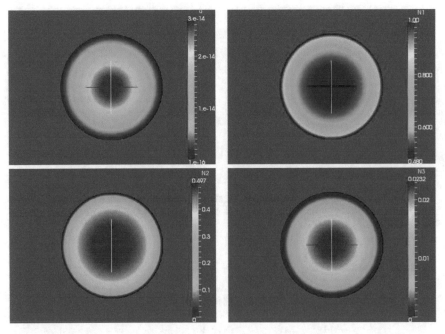

Fig. 4.16 Steady state solution of (4.65)–(4.68) with (4.82), (4.83), and the parameters in Table 4.3 (with permission from AIMS)

and N_3 the reverse is observed. The highest values are taken in the center of the domain and decrease toward the boundary. The density of swelling mitochondria attains in the center its maximum at around 0.497 and remains 0 in a layer along the boundary of Ω, where $u < C^-$ for all t. The density of fully swollen mitochondria N_3 nowhere attains values larger than 0.0232 due to a lack of u. Since $u < C^-$ always in a layer along the boundary, $N_3 = 0$ there for all t as swelling is not initiated. The radial gradients for N_1 and N_2 toward the boundary are comparable, whereas it is smaller for N_3 due to the lag. Overall N_3 is almost negligible compared to N_1 and N_2, wherefore $N_2 \approx 1 - N_1$.

In all cases, the numerical simulation maintains the radial symmetry of the problem, as it should.

In an additional simulation we tested the dependence of the steady state values on the lower induction threshold C^- that controls when the swelling starts locally in dependence of u. We varied this parameter between 0 and 15 and plot the evolution of the mitochondria densities averaged over Ω in Fig. 4.17. Comparing these data with each other and with Fig. 4.15d, which represents the case $C^- = 20$, reveals that the effect of this parameter is not very pronounced. This can be explained by the fast removal of u from the system through the boundaries, which leads to $u < C^-$ rapidly and thus puts a stop to the $N_1 \rightarrow N_2$ transition. Note that this statement likely holds for homogeneous Dirichlet problems only, as these parameters might play a crucial role in settings where the substrate concentration at the boundary is higher, e.g., in the case of nonhomogeneous Dirichlet conditions, Robin or Neumann conditions.

Discussion and Conclusion

The behavior of the mitochondrial swelling model for homogeneous Dirichlet boundary conditions is very different from the behavior previously observed for homogeneous Neumann boundary conditions.

In particular, these boundary conditions prevent complete swelling and lead to the system attaining a state in which some mitochondria close to the boundary of the domain are not engaged in swelling at all. Furthermore, in the interior of the domain mitochondria remain in the intermediate state but do not reach the complete swollen state. The reason for this behavior is that homogeneous Dirichlet conditions act as a calcium ion sink, leading to diffusive depletion of this substrate which controls the swelling process. For the swelling process to be initiated a certain minimum concentration is required. If the boundary value lies below these value (as in the homogeneous case), a boundary layer forms in which swelling is not initiated.

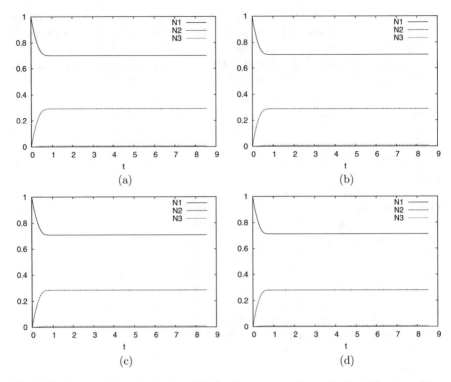

Fig. 4.17 Time evolution of mitochondrial fractions, averaged over Ω, for different swelling induction thresholds C^-. (**a**) $C^- = 0$. (**b**) $C^- = 5$. (**c**) $C^- = 10$. (**d**) $C^- = 15$ (with permission from AIMS)

References

1. A. Babin, *Attractors of Evolution Equations* (American Mathematical Society, Providence, 1992)
2. H. Brézis, *Opérateurs maximaux monotones et semi-groupes de contractions dans les espaces de Hilbert*, French. North-Holland Mathematics Studies. 5. Notas de matematica (50). Amsterdam-London: North-Holland Publishing Company; New York: American Elsevier Publishing Company, 183 p. Dfl. 25.00; ca. $ 8.80 (1973)
3. V.V. Chepyzhov, M.I. Vishik, *Attractors for Equations of Mathematical Physics*, vol. 49 (American Mathematical Society Providence, Providence, RI, 2002)
4. S. Eisenhofer, A coupled system of ordinary and partial differential equations modeling the swelling of mitochondria. PhD thesis, Technische Universität München, 2013
5. S. Eisenhofer, M.A. Efendiev, M. Otani, S. Schulz, H. Zischka, On a ODE–PDE coupling model of the mitochondrial swelling process, in Discrete Contin. Dyn. Syst. Ser. B **20**(4), 1031–1058 (2015)
6. S. Eisenhofer, F. Toókos, B.A. Hense, S. Schulz, F. Filbir, H. Zischka, A mathematical model of mitochondrial swelling. BMC Res. Not. **3**(1), 1 (2010)
7. M. Otani, Nonmonotone perturbations for nonlinear parabolic equations associated with subdifferential operators, Cauchy problems. J. Differ. Equ. **46**(2), 268–299 (1982)

8. A.V. Pokhilko, F.I. Ataullakhanov, E.L. Holmuhamedov, Mathematical model of mitochondrial ionic homeostasis: three modes of Ca 2+ transport. J. Theor. Biol. **243**(1), 152–169 (2006)
9. M. Renardy, R.C. Rogers, *An Introduction to Partial Differential Equations*, vol. 13 (Springer Science and Business Media, New York, 2006)
10. R. Temam, *Infinite-Dimensional Dynamical Systems in Mechanics and Physics.*, vol. 68 (Springer Science and Business Media, New York, 2012)
11. H. Triebel, *Interpolation Theory, Function Spaces, Differential Operators* (Leipzig, Barth, 1995), p. 532
12. E. Zeidler, *Functional Analysis and its Applications, I. Fixed Point Theorem* (Springer, New York, 1990)

Chapter 5
The Swelling of Mitochondria: In Vivo

In this chapter the swelling process is examined in a living organism, where we do not have a controlled environment as we had in the test tube. We take a look at the whole cell and analyze the effects of high Ca^{2+} concentrations to the mitochondria residing inside the cell. In contrast to the in vitro case, here calcium is not artificially added, but rather several biochemical processes in the organism lead to an increase of intracellular Ca^{2+}.

Remark 5.1 Ca^{2+} plays an important role in the communication of cells, not only in regard of inducing apoptosis by mitochondrial swelling. As it is, e.g., described in [15], calcium acts as a "second messenger," i.e., it is a chemical substance that translates extracellular signals into intracellular ones. This signal transduction by means of calcium ions is also involved in the activation of muscle contraction, cell division, or gene expression. Even for mitochondria, the main physiological role of Ca^{2+} uptake is control of the ATP production rate. Only the dose decides whether more energy is produced or the cell dies via apoptosis [13].

It is known that mitochondria within cells are not distributed randomly but reside in three main regions. As it was noted in [10], mitochondria reside around the nucleus, in a neighboring group and near the cell membrane. In liver cells, the total mitochondrial population comes up to 22% and the endoplasmic reticulum to 15% of the cell volume [1]. Figure 5.1 shows the organization of an eukaryotic cell, restricted to the cell compartments which are of interest for our purpose.
The endoplasmic reticulum is the main calcium storage inside a cell and plays a big role in the cellular Ca^{2+} homeostasis [13]. Another important feature are the calcium channels, which allow for an ion exchange between the cell and the extracellular regime.

It is not clear whether all mitochondria in a cell react in the same way, or if, e.g., mitochondria located around the nucleus are less sensitive to calcium than mitochondria residing near the cell membrane. At the moment this is not biologically clarified yet, but as we will see, we can introduce such kind of

© Springer Nature Switzerland AG 2018
M. Efendiev, *Mathematical Modeling of Mitochondrial Swelling*,
https://doi.org/10.1007/978-3-319-99100-9_5

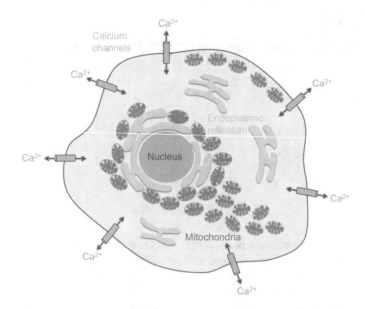

Fig. 5.1 Schematic diagram of a cell with focus on the mitochondrial distribution (with permission from AIMS)

properties into our mathematical model by assuming space dependencies of the mitochondrial behavior. Now it would be interesting to see if a given amount of calcium is sufficient to induce swelling for all mitochondria or if the process only attacks mitochondria from a specific region. At this the Ca^{2+} source plays a major role.

5.1   Increase of Intracellular Ca^{2+}

There are two mechanisms that lead to calcium increase inside the cell (see, e.g., [12, 13]), which both may lead to apoptosis, as it is depicted in Fig. 5.2.

- *Internal:* Ca^{2+} release from endoplasmic reticulum (ER)

  Due to external stimuli, Ca^{2+} is released from the endoplasmic reticulum. When ER stress is triggered by, e.g., the exposure to toxins or under pathophysiological conditions like ischemia or viral infections [14], this ER store is depleted and the released Ca^{2+} causes apoptotic events [4].
- *External:* Ca^{2+} influx from extracellular milieu

  The cell membrane itself is impermeable to ions and with that also to Ca^{2+}. However, there are transport systems that enable the calcium flux over the membrane. These are ion channels for the influx, whereas the efflux is controlled by Na^{+}/Ca^{2+} exchangers and ATP-dependent Ca^{2+} pumps. As we noted before, calcium acts as a second messenger and hence it is of major

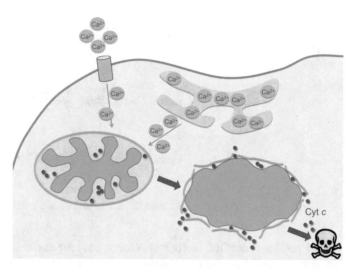

Fig. 5.2 Calcium triggers apoptosis either intra- or extracellularly (with permission from AIMS)

importance to maintain a constant cytosolic Ca^{2+} concentration, since a small change may cause severe cellular responses. For that reason there exists an enormous concentration gradient between cytosolic (100nM) and extracellular (1mM) calcium, which is to be maintained actively by use of energy [13].

If apoptosis now is induced by Ca^{2+} as an extracellular signal, it is exposed to the cell in the form of a directed calcium flow that enters the cell via the ion channels in the membrane [15].

For our model this increase of cytosolic Ca^{2+} signifies the start of the swelling process. It represents the initial calcium concentration by setting

$$u_0(x) := C_{in} + u_{peak}(x),$$

where C_{in} denotes the constant cytosolic concentration and $u_{peak}(x)$ describes the local increase. Here we assume this peak to be a singular event and model the resulting events effectuated from this calcium pulse.

A remarkable effect compared to the in vitro case is that for both possibilities of intracellular calcium increase the calcium source is very localized and by no means we have a uniformly distributed initial concentration.

5.2 The Cell Membrane

As described earlier, Ca^{2+} can enter or leave the cell across the plasma membrane via a nonsymmetric transport system consisting of channels and pumps. Here the radius and the location of these passage ways are not fixed and dependent on the present need of concentration gradient stabilization.

Fig. 5.3 Passing to the limit
of the ion channel radius
(with permission from AIMS)

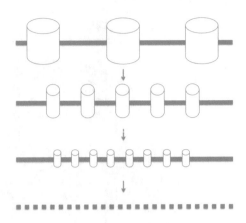

Since we do not have detailed biological information, for the mathematical modeling we assume that the total channel size over the whole membrane stays constant. Hence scaling down the radius implies a larger number of channels and passing to the limit as we did in Fig. 5.3 leads to an all-over permeable "limit membrane." By simplicity, in the following we will consider the in vivo model with this limit membrane.

Remark 5.2 In order to avoid this limiting process, we can introduce two regions Γ_1 and Γ_2 with $\Gamma_1 \cap \Gamma_2 = \emptyset$, where Γ_1 contains the channels and Γ_2 the closed parts of the membrane. Then by definition, on Γ_2 calcium cannot leave or enter the cell, which implies a different behavior on these two regions.

This poses difficulties for the mathematical analysis since we have to handle these two regions separately. However, it is not clear how to define the size of these subsets in a biological correct way and hence we cannot obtain a better description of the biological reality.

5.3   Summary

The previous observations make clear why it is really necessary to take into account spatial effects for the in vivo modeling of mitochondrial swelling. Briefly summarized, there are three main factors that differ a lot from the in vitro case:

- Mitochondria are not uniformly distributed.
- The inducing Ca^{2+} source is very localized.
- The cell is not a closed system.

5.4 Mathematical Analysis of an Vivo Model of Mitochondria Swelling

In this section, we further develop the model that we introduced in Chap. 4 and that takes into account the spatial effects. More precisely, two spatial effects directly influence the process of mitochondria swelling: on the one hand, the extent of mitochondrial damage due to calcium is highly dependent on the position of the particular mitochondrion and the local calcium ion concentration there. On the other hand, at a large amount of swollen mitochondria the effect of positive feedback becomes relevant as the residual mitochondria are confronted with a higher calcium ion load. This results in a coupled PDE-ODE system, see (5.1)–(5.4) below.

As we already mentioned in the previous sections in accordance with theoretical [8] and experimental [16] findings, we consider three subpopulations of mitochondria with different corresponding volumes: $N_1(x, t)$ describes the density of intact, unswollen mitochondria, $N_2(x, t)$ is the density of mitochondria that are in the swelling process but not completely swollen, and $N_3(x, t)$ is the density of completely swollen mitochondria. The swelling process is controlled by and affects the local Ca^{2+} concentration, which is denoted by $u(x, t)$, and subject to Fickian diffusion.

The transition of intact mitochondria over swelling to completely swollen ones proceeds in dependence on the local calcium ion concentration. Furthermore, as in the previous sections, we assume that mitochondria do not move in any direction and hence the spatial effects are only introduced by the calcium evolution. The evolution of the mitochondrial subpopulations is modeled by a system of ODEs that depends on the space variable x in terms of a parameter.

We analyze the swelling of mitochondria on a bounded domain $\Omega \subset \mathbb{R}^n$ with $n = 2, 3$. This domain is in this case the whole cell. The initial calcium concentration $u(x, 0)$ describes the added amount of Ca^{2+} to induce the swelling process. This leads to the following coupled PDE-ODE system determined by the nonnegative model functions f and g:

$$\partial_t u = d_1 \Delta_x u + d_2 g(u) N_2 \tag{5.1}$$

$$\partial_t N_1 = -f(u) N_1 \tag{5.2}$$

$$\partial_t N_2 = f(u) N_1 - g(u) N_2 \tag{5.3}$$

$$\partial_t N_3 = g(u) N_2 \tag{5.4}$$

with diffusion constant $d_1 > 0$ and feedback parameter $d_2 > 0$.

As we are interested in the in vivo case, we assume the boundary to be the permeable "limit membrane." Here calcium ions can enter or leave the cell over this membrane. The concentration gradient between the cell and the extracellular regime needs always be maintained, hence we assume inhomogeneous Robin boundary conditions

$$-\partial_\nu u(x,t) = a(x)\,(u(x,t) - \beta\,C_{ext}) \text{ for } x \in \partial\Omega\,.$$

Here $C_{ext} \geqslant 0$ denotes the constant extracellular calcium ion concentration and $\beta > 0$ represents the concentration gradient.

For instance, with the constants reported in [12], we have $C_{in} = 100\,\text{nM} = 10^{-7}\,\text{M}$ and $C_{ext} = 1\,\text{mM} = 10^{-3}\,\text{M}$, and hence the concentration gradient is of order 10^{-4} and we take $\beta = 10^{-4}$.

Remark 5.3

1) In general the extracellular calcium concentration is not constant; however, due to its largeness compared to the cell size, single calcium ion peaks are dissolved very fast.
2) By the choice of function a we can distinguish between different parts of the membrane. The previously mentioned case $\partial\Omega = \Gamma_1 \cup \Gamma_2$ hence could be realized by setting $a(x) = 0$ for $x \in \Gamma_2$ representing the closed parts of the membrane. This leads to zero flux on Γ_2 and concentration-dependent flux on Γ_1, just as we described the situation for the original membrane.
3) By the choice of function a we can switch between Dirichlet and Neumann type boundary conditions. If a is very small, the flux over the boundary is also very small and in the limit case $a \to 0$ we have homogeneous Neumann boundary conditions. On the other hand, for high values of a the solution soon approaches $u = \beta\,C_{ext}$ on the boundary, i.e., we can expect a behavior similar to nonhomogeneous Dirichlet boundary conditions.

The initial conditions are specified as

$$u(x,0) = u_0(x), \quad N_1(x,0) = N_{1,0}(x), \quad N_2(x,0) = N_{2,0}(x),$$
$$N_3(x,0) = N_{3,0}(x).$$

Note that by virtue of (5.2)–(5.4), the total mitochondrial population

$$\overline{N}(x,t) := N_1(x,t) + N_2(x,t) + N_3(x,t)$$

does not change in time, that is, $\partial_t \overline{N}(x,t) = 0$, and is given by the sum of the initial data:

$$\overline{N}(x,t) = \overline{N}(x) := N_{1,0}(x) + N_{2,0}(x) + N_{3,0}(x) \quad \forall t \geqslant 0 \ \forall x \in \Omega. \tag{5.5}$$

The role of model functions f and g as well as calcium evolution $u(x,t)$ is the same as was explained in detail in Sect. 4.1. We now give precise mathematical assumptions on f and g.

Condition 5.1 The model functions $f : \mathbb{R} \to \mathbb{R}$ and $g : \mathbb{R} \to \mathbb{R}$ have the following properties:

(i) Nonnegativity:

$$f(s) \geqslant 0 \qquad \forall s \in \mathbb{R},$$
$$g(s) \geqslant 0 \qquad \forall s \in \mathbb{R}.$$

(ii) Boundedness:

$$f(s) \leqslant f^* < \infty \qquad \forall s \in \mathbb{R},$$
$$g(s) \leqslant g^* < \infty \qquad \forall s \in \mathbb{R} \qquad \text{with } f^*, g^* > 0.$$

(iii) Lipschitz continuity:

$$|f(s_1) - f(s_2)| \leqslant L_f |s_1 - s_2| \qquad \forall s_1, s_2 \in \mathbb{R},$$
$$|g(s_1) - g(s_2)| \leqslant L_g |s_1 - s_2| \qquad \forall s_1, s_2 \in \mathbb{R} \qquad \text{with } L_f, L_g \geqslant 0.$$

In order to derive the uniform convergence of solutions, we need to introduce additional structure conditions on f and g. To do this, we distinguish between two cases, the cases $\alpha = 0$ and $\alpha > 0$. For the case $\alpha = 0$, we assume conditions similar to those for Dirichlet BC case and for $\alpha > 0$, similar to those for Neumann BC case.

Condition 5.2 (The case $\alpha = 0$) Let f and g fulfill Condition 5.1. In addition we assume that there exist constants $C^- > 0$, $m_1 > 0$, $m_2 > 0$, $\delta_0 > 0$, and $\varrho_0 > 0$ such that the following assertions hold:

(i) Starting threshold:

$$f(s) = 0 \qquad \forall s \leqslant C^-,$$
$$g(s) = 0 \qquad \forall s \leqslant 0.$$

(ii) Smoothness of f and g near starting threshold $[C^-, C^- + \delta_0]$ and $[0, \rho_0]$:

$$|f'(s)| \leqslant m_f s \qquad \forall s \in [C^-, C^- + \delta_0],$$
$$|g'(s)| \leqslant m_g s \qquad \forall s \in [0, \rho_0].$$

(iii) Lower bounds:

$$f(s) \geqslant f(C^- + \delta_0) > 0 \qquad \forall s \geqslant C^- + \delta_0,$$
$$g(s) \geqslant g(\varrho_0) > 0 \qquad \forall s \geqslant \varrho_0 > 0.$$

(iv) Dominance of g over f: There exists a constant $B > 0$ such that

$$f(s) \leqslant B g(s) \qquad \forall s \in [0, \infty).$$

Remark 5.4 It is easy to verify that Condition (5.2) is satisfied by monotone increasing functions f, $g \in C^2(\mathbb{R}^1)$ with $f(0) = g(0) = f'(C^-) = g'(0) = 0$.

Condition 5.3 (The case $\alpha > 0$) Let f and g fulfill Condition 5.1. We furthermore assume that there exist constants $C^- > 0$, $\delta_0 > 0$ and K_f such that the following assertions hold:

(i) Starting threshold:

$$f(s) = 0 \qquad \forall s \leqslant C^-,$$
$$g(s) = 0 \qquad \forall s \leqslant 0.$$

(ii) Smoothness of f near starting threshold $[C^-, C^- + \delta_0]$:

$$f(s) > 0 \qquad \forall s \in (C^-, C^- + \delta_0],$$

$$\frac{|f'(s)|^2}{f(s)} \leqslant K_f \qquad \forall s \in (C^-, C^- + \delta_0].$$

(iii) Lower bound:

$$f(s) \geqslant f(C^- + \delta_0) > 0 \qquad \forall s \geqslant C^- + \delta_0,$$

Remark 5.5 The above condition is similar to but weaker than Condition 2 assumed for Neumann BC case in [9], i.e.,

$$(C)_N \quad \exists m_1, m_2 \text{ such that} \qquad m_1(s - C^-) \leqslant f'(s)$$
$$\leqslant m_2(s - C^-) \qquad \forall s \in [C^-, C^- + \delta_0].$$

In fact, since this gives $m_1(s - C^-)^2/2 \leqslant f(s) \leqslant m_2(s - C^-)^2/2$, we easily have

$$\frac{|f'(s)|^2}{f(s)} \leqslant \frac{m_2^2(s - C^-)^2}{\frac{1}{2}m_1(s - C^-)^2} \leqslant \frac{2m_2^2}{m_1}.$$

The boundedness of $|\nabla_x N_2(t)|_{L^2(\Omega)}$ can be derived also for Neumann BCs with $(C)_N$ replaced by 2 of Condition 5.3 from much the simpler proof to be given later without distinguishing between two cases $u^\infty \leqslant C^-$ and $u^\infty > C^-$.

Well Posedness and Asymptotic Behavior of Solutions

For the analysis of (5.1)–(5.4), we write the Robin boundary condition in the form

$$-\partial_\nu u = a(u - \alpha) \text{ on } \partial\Omega. \tag{5.6}$$

The constant $\alpha \geqslant 0$ represents here the balance of concentration that is to be maintained. The boundary function $a(x)$ may be used to distinguish between different parts of the cell membrane. As is mentioned in Remark 5.3, we here allow that $a(x)$ can vanish somewhere on $\partial\Omega$:

$$a \in C^1(\partial\Omega), \quad 0 \leqslant a(x) \quad \text{for a.e. } x \in \partial\Omega \quad \text{and } a \not\equiv 0. \tag{5.7}$$

Remark 5.6 The assumption $a \in C^1(\partial\Omega)$ in (5.7) can be replaced by $a \in L^\infty(\partial\Omega)$ in the following arguments except in Proposition 5.2 and Theorem 5.3, where C^1-regularity of a is needed to assure the classical regularity of $\underline{v}$ and φ_1.

In the following we denote by (u, N_1, N_2, N_3) the corresponding solution of the Robin problem (5.1)–(5.4) with (5.6). Here many of the mathematical tools used in [5, 9] do not apply anymore and different arguments are required, see [6].

We state now our first main result:

Theorem 5.1 *Let $\Omega \subset \mathbb{R}^n$ be bounded. Assume Condition (5.1) and (5.7), then it holds:*

1. *For all initial data $u_0 \in L^2(\Omega)$ and $N_{i,0} \in L^\infty(\Omega)$ ($i = 1, 2, 3$), the system (5.1)–(5.4) with boundary condition (5.6) possesses a unique global solution (u, N_1, N_2, N_3) which satisfies $u \in C([0, T]; L^2(\Omega))$; $\sqrt{T} \, \partial_t u, \sqrt{T} \, \Delta_x u \in L^2([0, T]; L^2(\Omega))$; $N_i \in L^\infty([0, T]; L^\infty(\Omega))$ ($i = 1, 2, 3$) for all $T > 0$.*
2. *Assume further that u_0, $N_{1,0}$, $N_{2,0}$, and $N_{3,0} \geqslant 0$. Then the solution (u, N_1, N_2, N_3) preserves nonnegativity. Furthermore, N_1, N_2, N_3 are uniformly bounded in $\Omega \times [0, \infty)$.*
3. *We have the strong convergence results:*

$$N_1(t) \xrightarrow{t \to \infty} N_1^\infty \geqslant 0 \qquad\qquad \text{in } L^p(\Omega), \quad 1 \leqslant p < \infty, \tag{5.8}$$

$$N_2(t) \xrightarrow{t \to \infty} N_2^\infty \geqslant 0 \qquad\qquad \text{in } L^p(\Omega), \quad 1 \leqslant p < \infty, \tag{5.9}$$

$$N_3(t) \xrightarrow{t \to \infty} N_3^\infty \leqslant \|\overline{N}\|_{L^\infty(\Omega)} \text{ in } L^p(\Omega), \quad 1 \leqslant p < \infty, \tag{5.10}$$

$$u(t) \xrightarrow{t \to \infty} u^\infty \equiv \alpha \qquad\qquad \text{in } L^2(\Omega). \tag{5.11}$$

4. *Let $\alpha > 0$ and assume the following additional condition on g:*

Condition 5.4   *There exists $\rho_0 > 0$ such that $g(s)$ is strictly monotone increasing on $[0, \rho_0]$ and $g(\rho_0) \leqslant g(s)$ for all $\rho_0 \leqslant s$.*

Then we have $N_2^\infty \equiv 0$.

Proof **1. The existence of a unique global solution.**
      We put $\overline{u} = u - \alpha$, then $\overline{u}$ satisfies the boundary condition

$$-\partial_\nu \overline{u} = a\overline{u} \quad \text{on } \partial\Omega \tag{5.12}$$

and Eqs. (5.1)–(5.4) with u, f, g replaced by $\bar{u}$, $\bar{f}(v) = f(v + \alpha)$ and $\bar{g}(v) = g(v + \alpha)$ respectively. In what follows, we designate $\bar{u}$, $\bar{f}$, and $\bar{g}$ again by u, f, and g, if no confusion arises. Here we note that $\bar{f}$ and $\bar{g}$ also satisfy Condition 5.1.

Set

$$\varphi(u) := \begin{cases} \dfrac{1}{2} \displaystyle\int_\Omega |\nabla_x u|^2 \, dx + \dfrac{1}{2} \displaystyle\int_{\partial\Omega} a|u|^2 \, dS & \text{if } u \in H^1(\Omega) \\ +\infty & \text{if } u \in L^2(\Omega)\backslash H^1(\Omega). \end{cases}$$

Then it is easy to show that $\varphi : L^2(\Omega) \rightarrow [0, \infty]$ becomes a lower semi-continuous convex functional and its subdifferential $\partial\varphi$ (a notion of generalized Fréchet derivative, see Brézis [2, 3]) coincides with the self-adjoint operator A defined by

$$Au = -\Delta_x u \quad \text{with domain} \quad D(A) = \{u \in H^2(\Omega); \ -\partial_\nu u = a(x)u \text{ on } \partial\Omega\}. \tag{5.13}$$

In order to assure the local and global existence of a solution (u, N_1, N_2, N_3) to the original system, we can repeat the same arguments as for the Neumann boundary case, see [9]. We first note that by (5.5) the essential unknown functions can be taken as (u, N_1, N_2). Let $X_T := C([0, T]; L^2(\Omega))$ and define the mapping

$$\mathcal{B} : u \in X_T \mapsto N^u := (N_1^u, N_2^u) \mapsto \hat{u} = \mathcal{B}(u).$$

Here for a given $u \in X_T$, $N^u = (N_1^u, N_2^u)$ denotes the solution of the ODE problem:

$$\partial_t N^u = (-f(u)N_1^u, \ f(u)N_1^u - g(u)N_2^u) =: F^u(N^u),$$
$$N^u(x, 0) = (N_{1,0}(x), N_{2,0}(x)) \tag{5.14}$$

and $\hat{u}$ denotes the solution of the PDE problem:

$$\partial_t \hat{u} = d_1 \Delta_x \hat{u} + d_2 g(\hat{u})N_2^u, \quad \partial_\nu \hat{u} + a(x)\hat{u}|_{\partial\Omega} = 0, \quad \hat{u}(x, 0) = u_0(x), \tag{5.15}$$

which is reduced to the abstract problem in $H = L^2(\Omega)$:

$$\frac{d}{dt}\hat{u}(t) + d_1 \partial\varphi(\hat{u}(t)) + B(\hat{u}(t)) = 0, \quad \hat{u}(0) = u_0, \quad \text{with } B(u)(\cdot, t) = -d_2 g(u(\cdot, t)).$$

Since, by Condition 5.1 F^u is Lipschitz continuous from $Y = L^\infty(\Omega) \times L^\infty(\Omega)$ into itself, the Picard-Lindelöf theorem assures the existence of the unique global solution $N^u \in C([0, \infty); Y)$ of (5.14) for each $u \in X_T$. Furthermore, since the mapping $u \mapsto B(u) = g(u)N_2^u$ is Lipschitz continuous from $L^2(\Omega)$ into itself by Condition 5.1, the standard argument shows that (5.15) has the unique solution $\hat{u} \in C([0, \infty); L^2(\Omega))$ satisfying $\sqrt{T} \partial_t \hat{u}, \sqrt{T} \Delta_x \hat{u} \in L^2_{loc}((0, \infty); L^2(\Omega))$ (see, e.g., [2, 3, 11]).

Then, in view of the fact:

$$(\partial\varphi(u), u) = \int_\Omega |\nabla_x u|^2 dx + \int_{\partial\Omega} a|u|^2 dS \geq 0 \qquad \forall u \in D(A),$$

we can repeat exactly the same arguments as for the Neumann BC case in [9] and show that $\mathcal{B}$ becomes a contraction mapping in X_T for a sufficiently small $T_0 \in$ (0, 1] and $T_0 > 0$ does not depend on the choice of the initial data. Hence this local solution can be continued up to $[0, T]$ for any T.   □

In order to discuss the positivity and the asymptotic behavior of the solution, we need to first prepare some auxiliary results.

Proposition 5.1 (Comparison Theorem) *Let $d > 0$, $\alpha \in \mathbb{R}^1$, $a(x) \geq 0$, and $h :$ $\mathbb{R}^1 \to \mathbb{R}^1$ be Lipschitz continuous. Let $u_i \in \{u \in C([0, T]; L^2(\Omega); \sqrt{T}\partial_t u, \sqrt{T}\Delta_x u$ $\in L^2(0, T; L^2(\Omega))\}$ $(i = 1, 2)$ satisfy*

$$\begin{cases} \partial_t u_1 - d\Delta_x u_1 \geq h(u_1) \quad (x, t) \in \Omega \times (0, T), \quad -\partial_\nu u_1 \leq a(x)(u_1 - \alpha) \quad (x, t) \in \partial\Omega \times (0, T), \\ \partial_t u_2 - d\Delta_x u_2 \leq h(u_2) \quad (x, t) \in \Omega \times (0, T), \quad -\partial_\nu u_2 \geq a(x)(u_2 - \alpha) \quad (x, t) \in \partial\Omega \times (0, T), \\ u_1(x, 0) \geq u_2(x, 0) \quad x \in \Omega. \end{cases}$$

Then we have $u_1(x, t) \geq u_2(x, t)$ for a.e. $x \in \Omega$ and all $t > 0$.

Proof Let $w(t) = u_1(t) - u_2(t)$, then $w(t)$ satisfies

$$\partial_t w(t) - d\Delta_x w(t) \geq h(u_1(t)) - h(u_2(t)) \text{ in } \Omega, \quad -\partial_\nu w(t) \leq a(x)w(t)$$
$$\text{on } \partial\Omega, \quad w(0) \geq 0 \quad \text{in } \Omega.$$

Then multiplying this by $w^-(t) = \max(0, -w(t))$, we get

$$\frac{1}{2}\frac{d}{dt}\|w^-(t)\|^2_{L^2(\Omega)} + d\|\nabla_x w^-(t)\|^2_{L^2(\Omega)} + d\int_{\partial\Omega} \partial_\nu w\, w^- dS \leq L_h\|w^-(t)\|^2_{L^2(\Omega)},$$

where L_h is the Lipschitz constant of h. Noting that $\partial_\nu w\, w^- \geq -a w\, w^- = a|w^-|^2$ on $\partial\Omega$, we obtain

$$\frac{1}{2}\frac{d}{dt}\|w^-(t)\|^2_{L^2(\Omega)} + d\|\nabla_x w^-(t)\|^2_{L^2(\Omega)} + d\int_{\partial\Omega} a|w^-|^2 dS \leq L_h\|w^-(t)\|^2_{L^2(\Omega)}.$$

Then Gronwall's inequality implies that $|w^-(t)|^2 \leq |w^-(0)|^2 e^{L_h t} = 0$, whence follows $w^-(t) = 0$, i.e., $u_2(t) \leq u_1(t)$ for all $t > 0$.

Corollary 5.1 (Positivity) *Let $d > 0$, $\alpha \geq 0$ and $a(x) \geq 0$ and let u satisfy*

$$\partial_t u(t) - d\Delta_x u(t) \geq 0, \quad -\partial_\nu u(t) \leq a(x)(u(t) - \alpha), \quad u(0) \geq 0.$$

Then it holds that $u(x, t) \geq 0$ for a.e. $x \in \Omega$ and all $t > 0$.

Proof Since $\underline{u}(\cdot, t) \equiv 0$ satisfies

$$\partial_t \underline{u}(t) - d \Delta_x \underline{u}(t) = 0, \quad \underline{u}(\cdot, 0) = 0, \quad -\partial_\nu \underline{u}(t) = 0 \geqslant -\alpha a(x) = a(x)(\underline{u}(t) - \alpha),$$

Proposition 5.1 with $h \equiv 0$ assures that $u(x, t) \geqslant \underline{u}(x, t) \equiv 0$ for a.e. $x \in \Omega$ and all $t > 0$.

Proposition 5.2 (Strict Positivity) *Let* $d > 0$, $\alpha > 0$ *and* (5.7) *be satisfied. Suppose that u satisfies*

$$\begin{cases} \partial_t u - d \Delta_x u \geqslant 0 & (x, t) \in \Omega \times (0, T), \\ -\partial_\nu u \leqslant a(x)(u - \alpha) & (x, t) \in \partial\Omega \times (0, T), \\ u(x, 0) = u_0(x) \geqslant 0. \end{cases}$$

Then for any $t_* > 0$, *there exists* $\rho = \rho(t_*) > 0$ *such that*

$$u(x, t) \geqslant \rho \quad \text{for a.e. } x \in \Omega \text{ and all } t \geqslant t_*.$$

Proof Let $\underline{v}(\cdot, t)$ be the unique solution of

$$\partial_t \underline{v}(t) = d \Delta_x \underline{v}(t) \quad \text{in } \Omega, \quad -\partial_\nu \underline{v}(t) = a(x)(\underline{v}(t) - \alpha) \quad \text{on } \partial\Omega, \quad \underline{v}(0) = u_0 \quad \text{in } \Omega.$$

Then by Proposition 5.1 with $h \equiv 0$ and Corollary 5.1, we get $u(x, t) \geqslant \underline{v}(x, t) \geqslant 0$ for a.e. $x \in \Omega$ and all $t > 0$. Furthermore, by the strong parabolic maximum principle, we know $\underline{v}(x, t) > 0$ for all $(x, t) \in \Omega \times (0, T)$. For any $t_* > 0$, suppose that $\underline{v}(x_0, t_*) = 0$ for some $x_0 \in \partial\Omega$, then by virtue of Hopf's maximum principle, we get $\partial_\nu \underline{v}(x_0, t_*) < 0$, which implies

$$0 < -\partial_\nu \underline{v}(x_0, t_*) = a(x_0)(\underline{v}(x_0, t_*) - \alpha) = -a(x_0)\alpha \leqslant 0,$$

which leads to a contradiction. Hence there exists a positive constant $\rho_0 = \rho_0(t_*)$ such that

$$u(x, t_*) \geqslant \underline{v}(x, t_*) \geqslant \min_{x \in \bar{\Omega}} \underline{v}(x, t_*) = \rho_0 \quad \text{for a.e. } x \in \Omega.$$

Then putting $\rho = \rho(t_*) = \min(\rho_0(t_*), \alpha) > 0$, we get

$$(\partial_t - d \Delta_x)\rho = 0, \quad -\partial_\nu \rho = 0 \geqslant a(x)(\rho - \alpha), \quad u(x, t_*) \geqslant \rho.$$

Then it follows from Proposition 5.1 with $h \equiv 0$ that $u(x, t) \geqslant \rho$ for a.e. $x \in \Omega$ and all $t \geqslant t_*$.

Now we are in the position to continue our proof of Theorem 5.1.

Proof of Theorem 5.1 (continued)

2. Nonnegativity of solutions.   Multiplying the corresponding equations for
$N_i(x, t)$ by $N_i^-(x, t) = \max(-N_1(x, t), 0)$ and using assumptions $f(u) \geqslant$
0, $g(u) \geqslant 0$, we can deduce that $d\|N_i(t)\|_{L^2(\Omega)}/dt \leqslant 0$, whence follows
$\|N_i^-(t)\|_{L^2(\Omega)} \leqslant \|N_i^-(0)\|_{L^2(\Omega)} = 0$, i.e., $N_i(x, t) \geqslant 0$. So u satisfies $\partial_t u(t) -$
$d_1 \Delta_x u(t) = d_2 g(u(t)) N_2(t) \geqslant 0$, then Corollary 5.1 assures that $u(x, t) \geqslant 0$.
Furthermore the nonnegativity of $N_i(t)$ and the conservation law (5.5) imply the
uniform boundedness of N_i such that $0 \leqslant N_i(x, t) \leqslant \|\bar{N}\|_{L^\infty}$ $(i = 1, 2, 3)$ for a.e.
$x \in \Omega$ and all $t > 0$.

3. Convergence.   From (5.2) and the nonnegativity result it holds in the point-
wise sense

$$\partial_t N_1(x, t) = -f(u(x, t)) N_1(x, t) \leqslant 0 \quad \forall t \geqslant 0 \quad \text{a.e. } x \in \Omega.$$

Hence the sequence is nonincreasing and bounded below by 0, whence follows the
convergence

$$N_1(x, t) \xrightarrow{t \to \infty} N_1^\infty(x) \geqslant 0 \quad \text{a.e. } x \in \Omega.$$

Furthermore, by (5.5) we get

$$|N_1^\infty(x)| \leqslant \|\overline{N}\|_{L^\infty(\Omega)}, \quad N_1(x, t) = |N_1(x, t)| \leqslant \|\overline{N}\|_{L^\infty(\Omega)} \quad \text{a.e. } x \in \Omega, \ t \in (0, \infty).$$

Then by virtue of the Lebesgue dominated convergence theorem, we conclude that
$N_1(\cdot, t)$ converges to N_1^∞ strongly in $L^1(\Omega)$ as $t \to \infty$. Thus to deduce (5.8), it
suffices to use the relation

$$\|N_1(t) - N_1^\infty\|_{L^p(\Omega)}^p \leqslant \left(\|N_1(t)\|_{L^\infty(\Omega)} + \|N_1^\infty\|_{L^\infty(\Omega)}\right)^{p-1} \|N_1(t) - N_1^\infty\|_{L^1(\Omega)}.$$

As for $N_3(x, t)$, since (5.4) implies that $N_3(x, t)$ is monotone increasing, we can
repeat the same argument as above to get (5.10).

Combining (5.5) with (5.8) and (5.10), we can easily deduce (5.9).

Here we are going to show that if $\alpha > 0$ and Condition 5.4 is satisfied, then
$N_2^\infty(x) \equiv 0$. To this end, we first note that the integration of (5.4) on $(0, t)$ gives

$$0 \leqslant \int_0^t g(u(x, s)) N_2(x, s) dt = N_3(x, t) - N_3(x, 0) \leqslant \|\overline{N}\|_{L^\infty(\Omega)} \quad \forall t > 0.$$
$$\tag{5.16}$$

Here take any $t_* > 0$ and we put $\underline{\rho} = \underline{\rho}(t_*) := \min(\underline{\rho}(t_*), \rho_0)$, then Proposition 5.2
and Condition 5.4 assure

$$0 < g(\underline{\rho}) \leqslant g(u(t)) \quad \text{for a.e. } x \in \Omega \text{ and } t \geqslant t_*. \tag{5.17}$$

Then by (5.16) and (5.17), we get

$$g(\underline{\rho}) \int_{t_*}^{\infty} \|N_2(t)\|_{L^1(\Omega)} dt \leqslant \int_{\Omega} \int_0^t g(u(x,s)) N_2(x,s)\, ds\, dx \leqslant |\Omega|\, \|\overline{N}\|_{L^{\infty}(\Omega)}.$$

Hence there exists a sequence $\{t_k\}_{k\in\mathbb{N}}$ with $t_k \to \infty$ such that

$$\lim_{k \to \infty} \|N_2(t_k)\|_{L^1(\Omega)} = 0.$$

Then (5.9) implies

$$\lim_{k \to \infty} \int_{\Omega} N_2(x, t_k)\, dx = \int_{\Omega} N_2^{\infty}(x)\, dx = 0,$$

whence follows $N_2^{\infty}(x) \equiv 0$ for a.e. $x \in \Omega$.

In the following we are going to show that u converges to α strongly in $L^2(\Omega)$.

Here we can neither apply Wirtinger's nor Poincaré's inequality. Instead, we make use of the Poincaré-Friedrichs' inequality

$$\|w\|_{L^2(\Omega)}^2 \leqslant C_F \left(\|\nabla_x w\|_{L^2(\Omega)}^2 + \int_{\partial\Omega} a(x)|w|^2\, dS \right) \qquad \forall w \in H^1(\Omega). \qquad (5.18)$$

from Lemma 1.2.

We again put $\overline{u} = u - \alpha$ and $\overline{g}(v) = g(v + \alpha)$ and recall that $\overline{u}$ satisfies the homogeneous boundary condition (5.12) and Eq. (5.1) with g replaced by $\overline{g}$. Multiplying (5.1) by $\overline{u}$ and using (5.7), we get

$$\frac{1}{2}\frac{d}{dt}\|\overline{u}\|_{L^2(\Omega)}^2 + d_1 \int_{\Omega} |\nabla_x \overline{u}|^2\, dx + d_1 \int_{\partial\Omega} a(x)|\overline{u}|^2\, dS \leqslant d_2 \|\overline{g}(\overline{u}) N_2\|_{L^2(\Omega)} \|\overline{u}\|_{L^2(\Omega)}.$$
$$(5.19)$$

Then by (5.18), we obtain

$$\frac{1}{2}\frac{d}{dt}\|\overline{u}\|_{L^2(\Omega)}^2 + \frac{d_1}{C_F}\|\overline{u}\|_{L^2(\Omega)}^2 \leqslant d_2 \|\overline{g}(\overline{u}) N_2\|_{L^2(\Omega)} \|\overline{u}\|_{L^2(\Omega)}$$

and Young's inequality yields

$$\frac{1}{2}\frac{d}{dt}\|\overline{u}\|_{L^2(\Omega)}^2 + \frac{d_1}{2C_F}\|\overline{u}\|_{L^2(\Omega)}^2 \leqslant \frac{d_2^2 C_F}{2d_1}\|\overline{g}(\overline{u}) N_2\|_{L^2(\Omega)}^2. \qquad (5.20)$$

Here we note that (5.16) and the boundedness of $\|N_2(t)\|_{L^\infty}$ and g imply

$$\int_0^\infty \|\overline{g}(\overline{u}(t)) N_2(t)\|^2_{L^2(\Omega)} dt$$
$$\leq \int_0^\infty \int_\Omega g(u(x,s)) N_2(x,s) dx\, ds\, g^* \|\overline{N}\|_{L^\infty} \leq g^* |\Omega| \|\overline{N}\|^2_{L^\infty}. \qquad (5.21)$$

Then applying (ii) of Proposition 4 in [9](note that this result is stated for $y \in C^1([t_0, \infty))$, but it obviously holds also for $y(t) = \|\overline{u}(t)\|_{z^2} \in W^{1,1}(t_0, \infty)$ with $\gamma_0 = \frac{d_1}{2C_F}$, $a(t) = \frac{d_2^2 C_F}{2 d_1} \|\overline{g}(\overline{u}) N_2\|^2_{L^2(\Omega)}$ and $t_0 = 0$, we can deduce that $\|\overline{u}(t)\|_{L^2(\Omega)} \to 0$ as $t \to \infty$, which is equivalent to (5.11). $\qquad\square$

As for the a priori estimates for $\overline{u}(t)$, we can obtain more minute information. If fact, integrating (5.19) over $(0, \infty)$ and using Poincaré-Friedrichs' inequality and (5.21), we have

$$\int_0^\infty \|\overline{u}(t)\|^2_{H^1} dt \leq C_0, \qquad (5.22)$$

where C_0 is a general constant depending on $\|u_0\|_{L^2(\Omega)}$. Furthermore, multiplying (5.1) by $-\Delta_x \overline{u}$, we get

$$\frac{1}{2} \frac{d}{dt} \left(\|\nabla_x \overline{u}(t)\|^2_{L^2(\Omega)} + \int_{\partial\Omega} a(x) |\overline{u}(t)|^2 dS \right)$$
$$+ d_1 \|\Delta_x \overline{u}(t)\|^2_{L^2(\Omega)} \leq d_2 \|\overline{g}(\overline{u}) N_2\|_{L^2(\Omega)} \|\Delta_x \overline{u}(t)\|_{L^2(\Omega)}. \qquad (5.23)$$

Here by (5.18) we have

$$\|\Delta_x v\|_{L^2(\Omega)} \|v\|_{L^2(\Omega)} \geq (-\Delta_x v, v)_{L^2(\Omega)}$$
$$= 2\varphi(v)(t) \geq \sqrt{2\varphi(v)(t)} \frac{1}{\sqrt{C_F}} \|v\|_{L^2(\Omega)} \qquad \forall u \in H^1(\Omega). \qquad (5.24)$$

Hence combining (5.23) and (5.24), we obtain

$$\frac{d}{dt} \varphi(\overline{u})(t) + \frac{d_1}{4} \|\Delta_x \overline{u}(t)\|^2_{L^2(\Omega)} + \frac{d_1}{C_F} \varphi(\overline{u})(t) \leq \frac{d_2^2}{d_1} \|\overline{g}(\overline{u}(t)) N_2(t)\|^2_{L^2(\Omega)}. \qquad (5.25)$$

Since $\|\overline{g}(\overline{u}(t)) N_2(t)\|^2_{L^2(\Omega)} \in L^1(0, \infty)$ by (5.21) and $\overline{u}(\delta) \in H^1(\Omega)$ for some arbitrarily small $\delta > 0$ by (5.22), (ii) of Proposition 4 in [9] with $y(t) = \varphi(\overline{u})(t)$, $\gamma_0 = \frac{d_1}{2C_F}$, $a(t) = \frac{d_2^2}{d_1} \|\overline{g}(\overline{u}) N_2\|^2_{L^2(\Omega)}$ and $t_0 = \delta$, we can deduce that $\varphi(\overline{u})(t) \to 0$

as $t \to \infty$, which gives:

$$\sup_{t > \delta} \|\bar{u}(t)\|_{H^1} \leqslant C_0(\delta), \quad \|\nabla_x \bar{u}(t)\|_{L^2(\Omega)}^2 + \int_{\partial \Omega} a(x)|\bar{u}(t)|^2 dS \to 0 \quad \text{as } t \to \infty,$$
(5.26)

where $C_0(\delta)$ is a general constant depending only on $\|u_0\|_{L^2(\Omega)}$ and δ. Moreover, integrating (5.25) over $(0, \delta)$, we obtain

$$\int_{\delta}^{\infty} \|\Delta_x \bar{u}(t)\|_{L^2(\Omega)}^2 dt = \int_{\delta}^{\infty} \|\Delta_x u(t)\|_{L^2(\Omega)}^2 dt \leqslant C_0(\delta).$$
(5.27)

Uniform Convergence of u

In order to analyze more minute asymptotic behavior of solutions, we need the uniform convergence of u. For this purpose, we first prepare the following lemma.

Lemma 5.1 *Let Ω be a bounded domain in $\mathbb{R}^n$ with $n \leqslant 3$ and assume Conditions (5.1) and (5.7). Furthermore suppose that the following estimate holds:*

$$\sup_{t \geqslant 0} \|N_2(t)\|_{H^1} \leqslant C_{N_2} < \infty.$$
(5.28)

Then we have

$$\max_{x \in \overline{\Omega}} |u(x, t) - \alpha| \to 0 \quad as \ t \to \infty.$$
(5.29)

Proof We first note that $H^2(\Omega)$ is continuously embedded in the Hölder space $C^{\alpha}(\Omega)$ with order $\alpha \in (0, 1)$, since $n \leqslant 3$. Furthermore by virtue of the interpolation inequality, we get

$$\|u(t) - \alpha\|_{C^{\alpha}(\Omega)} \leqslant C_{\theta} \|u(t) - \alpha\|_{H^1}^{\theta} \|u(t) - \alpha\|_{H^2}^{1-\theta} \quad \theta \in (0, 1).$$

Hence, by virtue of (5.11) and (5.26), we find that in order to derive (5.29), it suffices to show that $\|\Delta_x u(t)\|_{L^2(\Omega)}$ is bounded on $[1, \infty)$. So we are going to show below that $\|\Delta_x u(t)\|_{L^2(\Omega)}$ is uniformly bounded. In what follows, we again denote $\bar{u}$, $\bar{g}$, $\bar{f}$ simply by u, g, f if no confusion arises. Here we recall the following facts on $A^{\frac{1}{2}}$, the fractional power of order $\frac{1}{2}$ of the operator A defined by (5.13):

$$D(A^{\frac{1}{2}}) = H^1(\Omega), \quad \|A^{\frac{1}{2}} u\|_{L^2(\Omega)}^2 = (Au, u) = \|\nabla_x u\|_{L^2(\Omega)}^2 + \int_{\partial \Omega} a|u|^2 dS.$$
(5.30)

Applying $A^{\frac{1}{2}}$ to (5.1), we consider the following auxiliary equation:

$$\partial_t A^{\frac{1}{2}} u + d_1 A^{\frac{3}{2}} u = d_2 A^{\frac{1}{2}} (g(u)N_2) .$$

By (5.22), (5.28), and (5.30), it is easy to see that $A^{\frac{1}{2}} (g(u)N_2) \in L^2_{loc}([0, \infty);$ $L^2(\Omega))$, then the standard regularity result assures that $\partial_t A^{\frac{1}{2}} u$ and $A^{\frac{3}{2}} u$ belong to $L^2_{loc}((0, \infty); L^2(\Omega))$. Then multiplying by $A^{\frac{3}{2}} u$, we have

$$\frac{1}{2} \frac{d}{dt} \|Au(t)\|^2_{L^2(\Omega)} + d_1 \|A^{\frac{3}{2}} u(t)\|^2_{L^2(\Omega)}$$
$$= d_2 \left(A^{\frac{1}{2}} (g(u(t))N_2(t)) , A^{\frac{3}{2}} u(t) \right)_{L^2(\Omega)} \qquad \text{for a.e. } t \in (0, \infty).$$

Hence by (5.30), we obtain

$$\frac{1}{2} \frac{d}{dt} \|\Delta_x u(t)\|^2_{L^2(\Omega)} + \frac{d_1}{2} \|A^{\frac{3}{2}} u(t)\|^2_{L^2(\Omega)} \leqslant \frac{d_2^2}{2 d_1} \|A^{\frac{1}{2}} (g(u(t))N_2(t)) \|^2_{L^2(\Omega)}$$

$$\leqslant \frac{d_2^2}{d_1} \left(\|g'(u(t))\nabla_x u(t)N_2(t)\|^2_{L^2(\Omega)} + \|g(u(t))\nabla_x N_2(t)\|^2_{L^2(\Omega)} \right.$$

$$\left. + \|a\|_{L^\infty} (g^*)^2 \|N_2(t)\|^2_{L^2(\partial\Omega)} \right)$$

$$\leqslant \frac{d_2^2}{d_1} \left(L_g^2 \|\overline{N}\|^2_{L^\infty} \|\nabla_x u(t)\|^2_{L^2(\Omega)} + (g^*)^2 \|\nabla_x N_2(t)\|^2_{L^2(\Omega)} \right.$$

$$\left. + C_\gamma \|a\|_{L^\infty} (g^*)^2 \|N_2(t)\|^2_{H^1} \right), \tag{5.31}$$

where C_γ is the embedding constant for $\|w\|^2_{L^2(\partial\Omega)} \leqslant C_\gamma \|w\|^2_{H^1(\Omega)}$.
Here we note that (5.18) and (5.30) yield $\|u\|^2_{L^2(\Omega)} \leqslant C_F \|A^{\frac{1}{2}} u\|^2_{L^2(\Omega)}$. Hence we get

$$\|A^{\frac{1}{2}} u\|^2_{L^2(\Omega)} = (A^{\frac{1}{2}} u, A^{\frac{1}{2}} u)_{L^2(\Omega)} = (Au, u)_{L^2(\Omega)} \leqslant \|Au\|_{L^2(\Omega)} \|u\|_{L^2(\Omega)}$$

$$\leqslant \|Au\|_{L^2(\Omega)} \sqrt{C_F} \|A^{\frac{1}{2}} u\|_{L^2(\Omega)}$$

$$\|A^{\frac{1}{2}} u\|^2_{L^2(\Omega)} \leqslant C_F \|Au\|^2_{L^2(\Omega)}$$

$$\|Au\|^2_{L^2(\Omega)} = (Au, Au)_{L^2(\Omega)} = (A^{\frac{1}{2}} u, A^{\frac{3}{2}} u)_{L^2(\Omega)} \leqslant \|A^{\frac{1}{2}} u\|_{L^2(\Omega)} \|A^{\frac{3}{2}} u\|_{L^2(\Omega)}$$

$$\leqslant \sqrt{C_F} \|Au\|_{L^2(\Omega)} \|A^{\frac{3}{2}} u\|_{L^2(\Omega)}$$

$$\|Au\|^2_{L^2(\Omega)} \leqslant C_F \|A^{\frac{3}{2}} u\|^2_{L^2(\Omega)}. \tag{5.32}$$

Hence by virtue of (5.31) and (5.32), we obtain

$$\frac{1}{2}\frac{d}{dt}\|\Delta_x u(t)\|^2_{L^2(\Omega)} + \frac{d_1}{2C_F}\|\Delta_x u(t)\|^2_{L^2(\Omega)} \leqslant a_A(t),$$

$$a_A(t) = \frac{d_2^2}{d_1}\left(L_g^2\|\overline{N}\|^2_{L^\infty}\|\nabla_x u(t)\|^2_{L^2(\Omega)} + (g^*)^2\|\nabla_x N_2(t)\|^2_{L^2(\Omega)} + C_\gamma\|a\|_{L^\infty}(g^*)^2\right.$$
$$\left.\|N_2(t)\|^2_{H^1}\right).$$

Here we recall that (5.26) and (5.28) assure that $\sup_{t\in[\delta,\infty)} a_A(t) \leqslant C_A(\delta)$ and that (5.27) implies that there exists $\delta_1 \in (\delta, 2\delta)$ such that $\|\Delta_x u(\delta_1)\|_{L^2(\Omega)} < \infty$. Thus applying (i) of Proposition 4 of [9] with $y(t) = \|\Delta_x u(t)\|^2_{L^2(\Omega)}$, $\gamma_0 = \frac{d_1}{2C_F}$, $a(t) \equiv C_A(\delta)$, $t_0 = \delta_1$ and $t_1 = \infty$, we can deduce

$$\sup_{t\in[\delta_1,\infty)}\|\Delta_x u(t)\|^2_{L^2(\Omega)} \leqslant \|\Delta_x u(\delta_1)\|^2_{L^2(\Omega)} + \frac{2C_F C_A(\delta)}{d_1},$$

whence follows the uniform boundedness of $\|\Delta_x u(t)\|_{L^2(\Omega)}$ on $[1, \infty)$. □

Boundedness of $\|\nabla_x N_{2(t)}\|_{L^2(\Omega)}$ Here we are going to show the boundedness of $\|\nabla_x N_2(t)\|_{L^2(\Omega)}$, for which we here assume the following:

Condition 5.5 $N_{1,0}, N_{2,0} \in H^1(\Omega)$.

(Case $\alpha > 0$) As for the case where $\alpha > 0$, our result is stated as follows.

Lemma 5.2 *Let $\alpha > 0$ and let all assumptions in Lemma 5.1 except (5.28) be satisfied. Further assume that Conditions 5.3, 5.4, and 5.5 are satisfied. Then there exists a constant C_N such that*

$$\sup_{t>0}\left(\|\nabla_x N_1(t)\|_{L^2(\Omega)} + \|\nabla_x N_2(t)\|_{L^2(\Omega)}\right) \leqslant C_N. \tag{5.33}$$

Proof Solving (5.2) point-wisely, we first get $N_1(x, t) = N_{1,0}(x)e^{-\int_0^t f(u(x,s))\,ds}$, which implies $N_1(t) \in H^1(\Omega)$ for all $t > 0$. Then applying the gradient to (5.2), we have

$$\partial_t \nabla_x N_1 = -f'(u)\nabla_x u N_1 - f(u)\nabla_x N_1. \tag{5.34}$$

Multiply (5.34) by $\nabla_x N_1(t)$, then we get

$$\frac{1}{2}\frac{d}{dt}\|\nabla_x N_1(t)\|^2_{L^2(\Omega)} \leqslant \int_\Omega |f'(u(x,t))||\nabla_x u(x,t)||N_1(x,t)||\nabla_x N_1(x,t)|\,dx \tag{5.35}$$
$$- \int_\Omega f(u(x,t))|\nabla_x N_1(x,t)|^2\,dx.$$

In view of Condition 5.3, we decompose Ω into 3 parts:

$$\Omega_1(t) := \{x \in \Omega; \, u(x,t) \leqslant C^-\},$$

$$\Omega_2(t) := \{x \in \Omega; \, C^- < u(x,t) \leqslant C^- + \delta_0]\}, \qquad (5.36)$$

$$\Omega_3(t) := \{x \in \Omega; \, C^- + \delta_0 < u(x,t)\}.$$

By virtue of Condition 5.3, we get

$$\int_\Omega |f'(u)||\nabla_x u||N_1||\nabla_x N_1|\, dx \leqslant \frac{1}{2}\int_{\Omega_2(t)\cup\Omega_3(t)} \frac{|f'(u)|^2}{f(u)}|N_1|^2|\nabla_x u|^2 dx$$

$$+ \frac{1}{2}\int_\Omega f(u)|\nabla_x N_1|^2 dx$$

$$\leqslant \frac{1}{2}\left(K_f\|\overline{N}\|_{L^\infty}^2 + \frac{L_f^2\|\overline{N}\|_{L^\infty}^2}{f(C^- + \delta_0)}\right)\|\nabla_x u\|_{L^2(\Omega)}^2 + \frac{1}{2}\int_\Omega f(u)|\nabla_x N_1|^2 dx.$$

Hence

$$\frac{1}{2}\frac{d}{dt}\|\nabla_x N_1(t)\|_{L^2(\Omega)}^2$$

$$\leqslant \frac{1}{2}\left(K_f\|\overline{N}\|_{L^\infty}^2 + \frac{L_f^2\|\overline{N}\|_{L^\infty}^2}{f(C^- + \delta_0)}\right)\|\nabla_x u(t)\|_{L^2(\Omega)}^2 \quad \text{for all } t > 0.$$

Then integrating this over $(0, t)$ and using (5.22), we deduce that there exists a constant C_{N_1} such that

$$\sup_{t>0}\|\nabla_x N_1(t)\|_{L^2(\Omega)} \leqslant C_{N_1}. \qquad (5.37)$$

By much the same argument as for (5.34), we get

$$\partial_t \nabla_x N_2 = f'(u)\nabla_x u\, N_1 + f(u)\nabla_x N_1 - g'(u)\nabla_x u\, N_2 - g(u)\nabla_x N_2. \qquad (5.38)$$

Multiplication of (5.38) by $\nabla_x N_2$ leads to

$$\frac{1}{2}\frac{d}{dt}\|\nabla_x N_2(t)\|_{L^2(\Omega)}^2 = \int_\Omega f'(u(t))\nabla_x u(t)N_1(t)\nabla_x N_2(t)\, dx$$

$$+ \int_\Omega f(u(t))\nabla_x N_1(t)\nabla_x N_2(t)\, dx$$

$$- \int_\Omega g'(u(t))\nabla_x u(t)N_2(t)\nabla_x N_2(t)\, dx$$

$$- \int_\Omega g(u(t))|\nabla_x N_2(t)|^2\, dx. \qquad (5.39)$$

Then Condition 5.1 yields

$$\frac{1}{2}\frac{d}{dt}\|\nabla_x N_2(t)\|^2_{L^2(\Omega)} \leqslant (L_f + L_g)\|\overline{N}\|_{L^\infty(\Omega)} \int_\Omega |\nabla_x u(t)||\nabla_x N_2(t)|\, dx$$

$$+ f^* \int_\Omega |\nabla_x N_1(t)||\nabla_x N_2(t)|\, dx - \int_\Omega g(u(t))|\nabla_x N_2(t)|^2\, dx \, .$$

Hence by Young's inequality and (5.37), for any $\eta > 0$, there exists a constant $C_{N_2}(\eta)$ such that

$$\frac{d}{dt}\|\nabla_x N_2(t)\|^2_{L^2(\Omega)} \leqslant \eta\|\nabla_x N_2(t)\|^2_{L^2(\Omega)} + C_{N_2}(\eta)\left(1 + \|\nabla_x u(t)\|^2_{L^2(\Omega)}\right)$$
$$\tag{5.40}$$
$$- 2\int_\Omega g(u(t))|\nabla_x N_2(t)|^2 dx.$$

Then (5.40) with $\eta = 1$ and Gronwall's inequality give

$$\|\nabla_x N_2(t)\|^2_{L^2(\Omega)}$$

$$\leqslant \left(\|\nabla_x N_{2,0}\|^2_{L^2(\Omega)} + C_{N_2}(1)\left(t_* + \int_0^{t_*} \|\nabla_x u(t)\|^2_{L^2(\Omega)}\, dt\right)\right)e^{t_*} \quad \forall t \in [0, t_*],$$
$$\tag{5.41}$$

where $t_* > 0$ is the number given in Proposition 5.2. As in the verification for $N_2^\infty(x) \equiv 0$, Proposition 5.2 and Condition 5.5 assure that

$$g(u(x,t)) \geqslant g(\underline{\rho}) > 0 \quad \forall t \geqslant t_* \tag{5.42}$$

with $\underline{\rho} = \underline{\rho}(t_*) = \min(\rho(t_*), \rho_0)$ as defined earlier. Substituting this into the last term of (5.40) with $\eta = g(\underline{\rho})$, we get

$$\frac{d}{dt}\|\nabla_x N_2(t)\|^2_{L^2(\Omega)} + g(\underline{\rho})\|\nabla_x N_2(t)\|^2_{L^2(\Omega)}$$

$$\leqslant C_{N_2}(g(\underline{\rho}))\left(1 + \|\nabla_x u(t)\|^2_{L^2(\Omega)}\right) \quad \forall t \geqslant t_*.$$

Then in view of (5.26), we can apply (i) of Proposition 4 in [9] with

$$y(t) = \|\nabla_x N_2(t)\|^2_{L^2(\Omega)}, \quad t_0 = t_*, \quad t_1 = \infty, \quad \gamma_0 = g(\underline{\rho}),$$

$$C = C_{N_2}(g(\underline{\rho}))\left(1 + \sup_{t \geqslant t_*} \|\nabla_x u(t)\|^2_{L^2(\Omega)}\right).$$

Thus together with (5.41), we achieved (5.33).

(**Case** $\alpha = 0$)   For the case where $\alpha = 0$, we cannot use the fact that $u(x, t)$ is bounded below by the positive constant ρ, so we here need to introduce more complicated arguments than before.   Our result for the case where $\alpha = 0$ is stated as follows.

Lemma 5.3 *Let $\alpha = 0$ and let all assumptions in Lemma 5.1 except (5.28) be satisfied. Further assume that Conditions 5.2 and 5.5 are satisfied. Then there exists a constant C_N such that*

$$\sup_{t>0} \left(\|\nabla_x N_1(t)\|_{L^2(\Omega)} + \|\nabla_x N_2(t)\|_{L^2(\Omega)} \right) \leqslant C_N. \tag{5.43}$$

Proof We define again $\Omega_i(t)$ $(i = 1, 2, 3)$ by (5.36), then in view of (5.35), by using $|f'(u)| \leqslant m_f |u|$ for $x \in \Omega_2(t)$, we have

$$\int_\Omega f'(u) \nabla_x u \, N_1 \nabla_x N_1 \, dx = \int_{\Omega_2(t)} f'(u) \nabla_x u \, N_1 \nabla_x N_1 \, dx$$

$$+ \int_{\Omega_3(t)} f'(u) \nabla_x u \, N_1 \nabla_x N_1 \, dx$$

$$\leqslant m_f \|N_1\|_{L^\infty(\Omega)} \int_{\Omega_2(t)} |u| \, |\nabla_x u| \, |\nabla_x N_1| \, dx + \|N_1\|_{L^\infty(\Omega)}$$

$$\int_{\Omega_3(t)} \frac{|f'(u)|}{\sqrt{f(u)}} |\nabla_x u| \sqrt{f(u)} |\nabla_x N_1| \, dx$$

$$\leqslant m_f \|N_1\|_{L^\infty(\Omega)} \int_\Omega |u| \, |\nabla_x u| \, |\nabla_x N_1| \, dx + \frac{L_f^2 \|N_1\|_{L^\infty(\Omega)}^2}{f(C^- + \delta_0)}$$

$$\int_\Omega |\nabla_x u|^2 \, dx + \frac{1}{4} \int_\Omega f(u) |\nabla_x N_1|^2 \, dx.$$

In view of (5.39), we put

$$J_1(t) := \int_\Omega |f'(u)| \, |\nabla_x u| \, |N_1| \, |\nabla_x N_2| \, dx, \quad J_2(t) := \int_\Omega f(u) |\nabla_x N_1| \, |\nabla_x N_2| \, dx,$$

$$J_3(t) := \int_\Omega |g'(u)| \, |\nabla_x u| \, |\nabla_x N_2| \, |N_2| \, dx, \quad J_4(t) := \int_\Omega g(u) |\nabla_x N_2|^2 \, dx.$$

It is obvious that for arbitrary $K > 0$ (which will be fixed later)

$$J_2(t) \leqslant \frac{K}{4} \int_\Omega f(u) |\nabla_x N_1|^2 \, dx + \frac{1}{K} \int_\Omega f(u) |\nabla_x N_2|^2 \, dx.$$

To estimate J_1 and J_3 we make the same trick as above:

$$|J_1(t)| \leqslant \int_{\Omega_2(t)} |f'(u)| |\nabla_x u| |\nabla_x N_2| |N_2| dx + \int_{\Omega_3(t)} |f'(u)| |\nabla_x u| |\nabla_x N_2| |N_2| dx$$

$$\leqslant m_f \|N_2\|_{L^\infty} \int_\Omega |u| |\nabla_x N_2| |\nabla_x u| dx$$

$$+ \|N_2\|_{L^\infty} \int_{\Omega_3(t)} \frac{f'(u)}{\sqrt{f(u)}} |\nabla_x u| \sqrt{f(u)} |\nabla_x N_2| dx$$

$$\leqslant m_f \|N_2\|_{L^\infty} \int_\Omega |u| |\nabla_x N_2| |\nabla_x u| dx$$

$$+ \frac{A^2 L_f^2 \|N_1\|_{L^\infty}^2}{f(C^- + \delta_0)} \int_\Omega |\nabla_x u|^2 dx + \frac{1}{4K} \int_\Omega f(u) |\nabla_x N_2|^2 dx.$$

$$|J_3(t)| \leqslant \int_\Omega |g'(u)| |\nabla_x u| |\nabla_x N_2| |N_2| \, dx$$

$$\leqslant \|N_2\|_{L^\infty(\Omega)} \int_{\{u \leqslant \rho_0\}} |g'(u)| |\nabla_x u| |\nabla_x N_2| dx$$

$$+ \|N_2\|_{L^\infty(\Omega)} \int_{\{u > \rho_0\}} |g'(u)| |\nabla_x u| |\nabla_x N_2| dx$$

$$\leqslant m_g \|N_2\|_{L^\infty(\Omega)} \int_{\{u \leqslant \rho_0\}} |u| |\nabla_x u| |\nabla_x N_2| dx$$

$$+ \|N_2\|_{L^\infty(\Omega)} \int_{\{u > \rho_0\}} \frac{g'(u)}{\sqrt{g(u)}} |\nabla_x u| \sqrt{g(u)} |\nabla_x N_2| dx$$

$$\leqslant m_g \|N_2\|_{L^\infty(\Omega)} \int_\Omega |u| |\nabla_x u| |\nabla_x N_2| dx$$

$$+ \frac{L_g^2 \|N_2\|_{L^\infty(\Omega)}^2}{g(\rho_0)} \int_\Omega |\nabla_x u|^2 dx + \frac{1}{4} \int_\Omega g(u) |\nabla_x N_2|^2 dx.$$

Hence there exists a constant C_1 such that

$$\partial_t \|\nabla_x N_1(t)\|_{L^2(\Omega)}^2 \leqslant C_1 \int_\Omega |u| |\nabla_x u| |\nabla_x N_1| dx + C_1 \int_\Omega |\nabla_x u|^2 dx$$
$$- \frac{3}{4} \int_\Omega f(u) |\nabla_x N_1|^2 dx,$$

$$\partial_t \|\nabla_x N_2(t)\|_{L^2(\Omega)}^2 \leqslant C_1 \int_\Omega |u| |\nabla_x u| |\nabla_x N_2| dx + C_1 \int_\Omega |\nabla_x u|^2 dx$$

$$+ \frac{K}{4} \int_\Omega f(u) |\nabla_x N_1|^2 dx$$

$$+ \frac{5}{4K} \int_\Omega f(u) |\nabla_x N_2|^2 dx - \frac{3}{4} \int_\Omega g(u) |\nabla_x N_2|^2 dx.$$

Let $y(t) := K \|\nabla_x N_1(t)\|^2_{L^2(\Omega)} + \|\nabla_x N_2(t)\|^2_{L^2(\Omega)}$, then we have

$$\partial_t y(t) \leqslant C_1 \int_\Omega |u| |\nabla_x u| (K |\nabla_x N_1| + |\nabla_x N_2|) dx + C_1 (K+1) \int_\Omega |\nabla_x u|^2 dx$$

$$- \frac{3K}{4} \int_\Omega f(u) |\nabla_x N_1|^2 dx + \frac{K}{4} \int_\Omega f(u) |\nabla_x N_1|^2 dx$$

$$+ \frac{5}{4K} \int_\Omega f(u) |\nabla_x N_2|^2 dx - \frac{3}{4} \int_\Omega g(u) |\nabla_x N_2|^2 dx.$$

Choosing $K = 2B$ (B is given in Condition 5.2), we obtain

$$\partial_t y(t) \leqslant C_1 \int_\Omega |u| |\nabla_x u| (K |\nabla_x N_1| + |\nabla_x N_2|) dx + C_1 (K+1) \int_\Omega |\nabla_x u|^2 dx$$

$$\leqslant 2 C_2 \|u\|_{L^4} \|\nabla_x u\|_{L^4} (K \|\nabla_x N_1\|^2_{L^2(\Omega)} + \|\nabla_x N_2\|^2_{L^2(\Omega)})^{\frac{1}{2}}$$

$$+ C_1 (K+1) \|\nabla_x u\|^2_{L^2(\Omega)}$$

$$\leqslant 2 C_2 \|u\|_{H^1} \|u\|_{H^2} (K \|\nabla_x N_1\|^2_{L^2(\Omega)} + \|\nabla_x N_2\|^2_{L^2(\Omega)})^{\frac{1}{2}}$$

$$+ C_1 (K+1) \|\nabla_x u\|^2_{L^2(\Omega)}$$

$$\leqslant C_2 \|u\|^2_{H^1} (K \|\nabla_x N_1\|^2_{L^2(\Omega)} + \|\nabla_x N_2\|^2_{L^2(\Omega)})$$

$$+ C_2 \|u\|^2_{H^2} + C_1 (K+1) \|\nabla_x u\|^2_{L^2(\Omega)}$$

$$\leqslant C_2 \|u\|^2_{H^1} y(t) + C_2 \|u\|^2_{H^2} + C_1 (K+1) \|\nabla_x u\|^2_{L^2(\Omega)},$$

where C_2 is a constant depending on C_1 and some embedding constants. Integrating the last inequality over (δ, t), we obtain

$$y(t) \leqslant y(\delta) + \int_\delta^T C_2 \|u\|^2_{H^1} y(s) ds + \int_\delta^T \left(C_2 \|u\|^2_{H^2} + C_1 (K+1) \|\nabla_x u\|^2_{L^2(\Omega)} \right) ds.$$

Hence, the Gronwall inequality leads to

$$y(t) \leqslant \left[y(\delta) + C_2 \int_\delta^\infty \|u\|_{H^2}^2 \, ds + C_1 (K+1) \int_0^\infty \|\nabla_x u\|_{L^2(\Omega)}^2 \, ds \right]$$
$$\exp \left(\int_0^\infty C_2 \|u\|_{H^1}^2 \, ds \right). \tag{5.44}$$

Here in view of (5.35) and (5.39), we get

$$\frac{d}{dt} \|\nabla_x N_1(t)\|_{L^2(\Omega)} \leqslant L_f \|\overline{N}\|_{L^\infty} \|\nabla_x u\|_{L^2(\Omega)},$$

$$\frac{d}{dt} \|\nabla_x N_2(t)\|_{L^2(\Omega)} \leqslant (L_f + L_g) \|\overline{N}\|_{L^\infty} \|\nabla_x u\|_{L^2(\Omega)} + f^* \|\nabla_x N_1(t)\|_{L^2(\Omega)}.$$

Integrating these inequalities over $(0, \delta)$ with respect to t and using the fact that $N_{1,0}, N_{2,0} \in H^1(\Omega)$ and (5.22), we can deduce the a priori bound for $y(\delta)$. Thus (5.43) is derived from (5.44) together with (5.22) and (5.27).   □

Next, we study partial and complete swelling scenarios of mitochondria in the case of inhomogeneous Robin boundary condition (5.6). We already saw in Chap. 4 that such scenarios crucially depend upon on the boundary conditions thus leading to different behavior of solutions in the case of Neumann and Dirichlet boundary conditions.

Partial Swelling

We first consider the case where $\alpha \in (0, C^-)$.

Case 1. $0 < \alpha < C^-$
    For this case, it occurs the partial swelling. More precisely we have:

Theorem 5.2 *Let all assumptions in Lemma 5.2 be satisfied and let $\alpha \in (0, C^-)$. Then there exists a finite time $T_p > 0$ such that*

$$N_1(x, t) = N_1(x, T_p) \quad \forall t \geqslant T_p, \tag{5.45}$$

and the following exponential convergences hold.

$$N_2(x, t) \xrightarrow{t \to \infty} 0 \qquad\qquad in\ \mathcal{O}(e^{-g(\underline{\rho})t}) \quad for\ a.e.\ x \in \Omega,$$

$$N_3(x, t) \xrightarrow{t \to \infty} \overline{N}(x) - N_1(x, T_p) \qquad in\ \mathcal{O}(e^{-g(\underline{\rho})t}) \quad for\ a.e.\ x \in \Omega,$$

$$\|u(t) - \alpha\|_{L^2(\Omega)}^2 \xrightarrow{t \to \infty} 0 \qquad\qquad in\ \mathcal{O}(e^{-\gamma_1 t}),$$

$$\|\nabla_x u(t)\|_{L^2(\Omega)}^2 \xrightarrow{t \to \infty} 0 \qquad\qquad in\ \mathcal{O}(e^{-\gamma_1 t}),$$

where $\rho = \rho(T_p) = \min(\rho(T_p), \rho_0)$ with ρ earlier given in Proposition 5.2 and γ_1 is any number satisfying $0 < \gamma_1 < \min\left(\frac{d_1}{C_F}, 2\,g(\underline{\rho})\right)$. Here the terminology $v(t) \xrightarrow{t \to \infty} v^\infty$ in $\mathcal{O}(e^{-kt})$ means that there exists some constant $C > 0$ such that

$$|v(t) - v^\infty| \leqslant C e^{-kt} \quad \text{for all } t \geqslant T_p.$$

Proof By virtue of Lemmas 5.1 and 5.2, we find that u converges to $\alpha < C^-$ uniformly. Then there exists a finite time T_p such that $u(x, t) \leqslant C^-$ for all $t \geqslant T_p$, which together with (i) of Condition 5.3 implies $f(u(x, t)) \equiv 0$ for all $t \geqslant T_p$. Then by (5.2), we get $N_1(x, t) = N_1(x, T_p) \; \forall t \geqslant T_p$. Hence by (5.3) and (5.17) (or (5.42)), we get

$$\partial_t N_2(x, t) = -g(u(x, t))\,N_2(x, t) \leqslant -g(\underline{\rho})\,N_2(x, t) \quad t \geqslant T_p,$$

$$N_2(x, t) \leqslant N_2(x, T_p) \exp(-g(\underline{\rho})(t - T_p)) \quad \forall t \in [T_p, \infty), \tag{5.46}$$

where $\underline{\rho} = \rho(T_p) = \min(\rho(T_p), \rho_0) > 0$ and ρ is given in (5.17) and Proposition 5.2. In order to see the exponential convergence of $N_3(x, t)$, it suffices to recall the conservation law (5.5) for $\overline{N}(x, t)$.

As for the convergence of $\overline{u}(t) = u(t) - \alpha$, by putting $\tilde{\gamma}_0 = \frac{d_1}{C_F}$, we obtain by (5.20) and (5.46)

$$\frac{d}{dt}\|\overline{u}(t)\|^2_{L^2(\Omega)} + \tilde{\gamma}_0\|\overline{u}\|^2_{L^2(\Omega)} \leqslant \frac{d_2^2}{\tilde{\gamma}_0}\|\overline{g}(u)\,N_2\|^2_{L^2(\Omega)}$$

$$\leqslant \frac{d_2^2}{\tilde{\gamma}_0}(g^*)^2\|\overline{N}\|^2_{L^\infty(\Omega)}|\Omega|^2 e^{2g(\underline{\rho})T_p} e^{-2g(\underline{\rho})t} \quad \forall t \geqslant T_p. \tag{5.47}$$

Here let γ_1 be any number satisfying

$$0 < \gamma_1 < \min\left(\frac{d_1}{C_F}, 2g(\underline{\rho})\right).$$

Then by (5.47), we get

$$\partial_t\left(e^{\gamma_1 t}\|\overline{u}(t)\|^2_{L^2(\Omega)}\right) \leqslant C_\gamma e^{-(2g(\underline{\rho}) - \gamma_1)t} \quad \forall t \geqslant T_p,$$

$$C_\gamma = \frac{d_2^2}{\tilde{\gamma}_0}(g^*)^2\|\overline{N}\|^2_{L^\infty(\Omega)}|\Omega|^2 e^{2g(\underline{\rho})T_p}.$$

Hence integrating this over (T_p, t), we obtain the exponential decay of $\|\overline{u}(t)\|^2_{L^2(\Omega)}$:

$$\|\overline{u}(t)\|^2_{L^2(\Omega)} \leqslant \left(e^{\gamma_1 T_p} \sup_{s>0}\|\overline{u}(s)\|^2_{L^2(\Omega)} + \frac{C_\gamma}{2g(\underline{\rho}) - \gamma_1} e^{-(2g(\underline{\rho}) - \gamma_1)T_p}\right) e^{-\gamma_1 t},$$

which implies

$$\|\overline{u}(t)\|^2_{L^2(\Omega)} \xrightarrow{t \to \infty} 0 \text{ in } \mathcal{O}(e^{-\gamma_1 t}) \quad \text{for any } \gamma_1 \in (0, \min{(d_1/C_F, 2g(\underline{\rho})))} .$$

As for $\varphi(\overline{u}(t))$, from (5.25) we now get

$$\frac{d}{dt}\varphi(\overline{u})(t) + \frac{d_1}{C_F}\varphi(\overline{u})(t) \leqslant \frac{d_2^2}{d_1}(g^*)^2\|N_2(t)\|^2_{L^2(\Omega)} .$$

Thus repeating the same argument as for $\|\overline{u}(t)\|^2_{L^2(\Omega)}$, we can obtain the convergence

$$\|\overline{u}(t)\|^2_{H^1} = \|\nabla_x u(t)\|^2_{L^2(\Omega)} + \|u(t) - \alpha\|^2_{L^2(\Omega)} \xrightarrow{t \to \infty} 0 \quad \text{in } \mathcal{O}(e^{-\gamma_1 t}) .$$

Case 2. $\alpha = 0$

We next consider the case where $\alpha = 0$. For this case, partial swelling also occurs.

However, the asymptotic behavior of $N_2(x, t)$ is quite different from that of the previous case.

Theorem 5.3 *Let all assumptions in Lemma 5.3 be satisfied. Then there exists a finite time $T_p > 0$ such that (5.45) holds. Furthermore, if $g(s)$ is monotone increasing in $[0, \rho_0]$ (ρ_0 is the parameter given in Condition 5.2) and if there exist $T_1 \in [0, \infty)$ and $\rho_1 > 0$ such that $N_2(x, T_1) \geqslant \rho_1$ for a.e. $x \in \Omega$, then there exists $\rho_2 > 0$ such that*

$$N_2^\infty(x) \geqslant \rho_2 \quad a.e. \ x \in \Omega.$$

Proof The first part can be derived from Lemmas 5.1 and 5.3 as before. To establish the positive lower bound for $N_2^\infty(x)$, we are going to construct a sub-solution for $N_2(x, t)$. To this end, we first construct a super-solution for $u(x, t)$ by making use of the first eigenfunction of the following eigenvalue problem:

$$(E)_\lambda \begin{cases} -\Delta_x \varphi = \lambda \varphi & \text{in } \Omega \\ -\partial_\nu \varphi = a(x)\varphi & \text{on } \partial\Omega. \end{cases}$$

Let

$$R(\varphi) := \frac{\int_\Omega |\nabla_x \varphi|^2 dx + \int_\Omega a(x)|\varphi(x)|^2 dx}{\int_\Omega |\varphi(x)|^2 dx}.$$

Using the standard compactness argument, one can easily prove that there exists a global minimizer φ_1 of R in $H^1(\Omega)$. Since $R(|\varphi|) = R(\varphi)$, without loss of generality we can take φ_1 such that $\varphi_1 \geqslant 0$, i.e., φ_1 satisfies $R(\varphi_1) = \inf_{v \in H^1(\Omega)} R(v) = \lambda_1 > 0$ and λ_1 is the first eigenvalue and φ_1 the first eigenfunction for $(E)_\lambda$. Here we normalize the first eigenfunction such that $\max_{x \in \overline{\Omega}} \varphi_1(x) = 1$. From the strong maximum principle, it follows that $\varphi_1(x) > 0$ in Ω. Moreover it holds that $\varphi_1(x) > 0$ in $\overline{\Omega}$. Indeed, assume on the contrary that there exists $x_0 \in \partial\Omega$ such that $\varphi_1(x_0) = 0$, then Hopf's strong maximum principle assures $\partial_\nu \varphi_1(x_0) < 0$, which contradicts the boundary condition $\partial_\nu \varphi_1(x_0) = -a(x)\varphi_1(x_0) = 0$. Hence, $\varphi_1(x) > 0$ on $\overline{\Omega}$ and there exists C^* such that $\min_{x \in \overline{\Omega}} \varphi_1(x) \geqslant \frac{C^*}{\rho_0} > 0$, where ρ_0 is the same as in Condition 5.2. Since $\|u(t)\|_{L^\infty(\Omega)} \to 0$ as $t \to \infty$, there exists $T_0 > 0$ such that

$$d_2 m_g \|N_2\|_{L^\infty(\Omega)} \|u(t)\|_{L^\infty(\Omega)} \leqslant d_1 \lambda_1,$$

$$\|u(t)\|_{L^\infty(\Omega)} \leqslant \min\left(C^*, \rho_0, C^-\right) \quad \text{for all } t \geqslant T_0.$$

Then due to the inequality $|g'(s))| \leqslant m_g |s|$ for all $|u| \leqslant \rho_0$ in Condition 5.2, we have

$$|d_2 g(u(t)) N_2(t)| \leqslant d_2 \frac{1}{2} m_g |u(t)|^2 \|N_2(t)\|_{L^\infty(\Omega)} \leqslant \frac{1}{2} d_1 \lambda_1 |u(t)| \quad \text{for all } t \geqslant T_0. \tag{5.48}$$

Let $\overline{u}(x, t) := \lambda(t)\varphi_1(x)$ and $\lambda(t)$ be the solution of

$$\begin{cases} \lambda'(t) + \dfrac{1}{2} d_1 \lambda_1 \lambda(t) = 0 & \text{for } t \geqslant T_0, \\ \lambda(T_0) = \rho_0, \end{cases}$$

more explicitly

$$\lambda(t) = \rho_0 \exp\left(-\frac{1}{2} d_1 \lambda_1 (t - T_0)\right) \quad \text{for } t \geqslant T_0.$$

Then

$$|\overline{u}(x, t)| = |\lambda(t)\varphi_1(x)| \leqslant \rho_0 \quad \text{for } t \geqslant T_0$$

and $\overline{u}(x, t)$ satisfies

$$\begin{cases} \partial_t \overline{u}(x, t) = d_1 \Delta_x \overline{u}(x, t) + \dfrac{1}{2} d_1 \lambda_1 \overline{u}(x, t) & (x, t) \in \Omega \times (T_0, \infty), \\ \overline{u}(x, T_0) = \rho_0 \varphi_1(x) \quad x \in \Omega, \quad -\partial_\nu \overline{u} = a(x)\overline{u}(x) & (x, t) \in \partial\Omega \times (T_0, \infty). \end{cases}$$

Here by (5.48), we note

$$\partial_t u - d_1 \Delta_x u = d_2 g(u) N_2(x,t) \leqslant \frac{1}{2} d_1 \lambda_1 u(x,t) \quad \text{for a.e. } x \in \Omega \text{ and all } t \geqslant T_0,$$

$$u(x, T_0) \leqslant C^* = \frac{C^*}{\rho_0} \cdot \rho_0 \leqslant \varphi_1(x) \cdot \lambda(T_0) = \bar{u}(x, T_0).$$

Thus by the comparison theorem Proposition 5.1 with $h(u) = \frac{1}{2} d_1 \lambda_1 u$ and with 0 replaced by T_0, we have

$$u(x,t) \leqslant \bar{u}(x,t) \leqslant \lambda(t) \leqslant \rho_0 e^{-\frac{1}{2} d_1 \lambda_1 (t-T_0)} \; \forall t \geqslant T_0 \text{ for a.e. } x \in \Omega \text{ and all } t \geqslant T_0.$$

Since $g(s)$ is monotone increasing on $[0, \rho_0]$, by the 2nd property of Condition 5.2, we obtain

$$g(u(x,t)) \leqslant g(\rho_0 e^{-\frac{1}{2} d_1 \lambda_1 (t-T_0)}) \leqslant \frac{1}{2} m_g \rho_0^2 e^{-d_1 \lambda_1 (t-T_0)}$$

for a.e. $x \in \Omega$ and all $t \geqslant T_0$.

Moreover, taking into account that $\|u(x,t)\|_{L^\infty(\Omega)} \leqslant C^-$ for $t \geqslant T_0$, we have $f(u(x,t)) = 0$ and consequently

$$\partial_t N_2(x,t) = -g(u(x,t)) N_2(x,t).$$

Thus $\partial_t N_2(x,t) \geqslant -\frac{1}{2} m_g \rho_0^2 e^{-d_1 \lambda_1 (t-T_0)} N_2(x,t)$ for all $t \geqslant T_0$, whence follows

$$\int_{T_0}^T \frac{dN_2}{N_2} \geqslant -\int_{T_0}^T \frac{1}{2} m_g \rho_0^2 e^{-d_1 \lambda_1 (t-T_0)} \, dt.$$

Hence

$$\log \frac{N_2(x,t)}{N_2(x,T_0)} \geqslant \left[\frac{1}{2} \frac{1}{d_1 \lambda_1} m_g \rho_0^2 e^{-d_1 \lambda_1 (s-T_0)} \right]_{s=T_0}^{s=t} = \frac{m_g \rho_0^2}{2 d_1 \lambda_1} \left[e^{-d_1 \lambda_1 (t-T_0)} - 1 \right].$$

Since $\partial_t N_2(x,t) < 0$ for $t \geqslant T_0$ we have

$$\frac{N_2(x,t)}{N_2(x,T_0)} < 1 \quad \text{and} \quad \log \frac{N_2(x,t)}{N_2(x,T_0)} < 0.$$

Consequently,

$$\frac{N_2(x,T_0)}{N_2(x,t)} \leqslant e^{\frac{m_g \rho_0^2}{2 d_1 \lambda_1} [1 - e^{-d_1 \lambda_1 (t-T_0)}]}$$

and letting $t \to \infty$, we get

$$N_2^\infty(x) \geq N_2(x, T_0) e^{\frac{mg\rho_0^2}{2d_1\lambda_1}}.$$

Then to complete the proof, it suffices to show that $\inf_{x\in\Omega} N_2(x, T_0) > 0$. For the case where $T_0 \leq T_1$, it is clear that we can repeat the same argument above with T_0 replaced by T_1. Hence the conclusion is obvious. As for the case where $T_1 < T_0$, we note that N_2 satisfies

$$\partial_t N_2(x, t) \geq -g^* N_2(x, t) \quad \text{for all } t > 0,$$

which implies

$$\partial_t \left[N_2(x, t) e^{g^*(t-T_1)} \right] \geq 0 \quad \Rightarrow \quad N_2(x, t) e^{g^*(t-T_1)} \geq N_2(x, T_1).$$

Thus we obtain

$$N_2(x, T_0) \geq e^{-g^*(T_0-T_1)} N_2(x, T_1) \geq e^{-g^*(T_0-T_1)} \rho_1 > 0 \quad \text{for a.e. } x \in \Omega.$$

Remark 5.7 The following two assumptions are sufficient conditions for $N_2(x, T_1) \geq \rho_1 > 0$ for a.e. $x \in \Omega$ with $T_1 \in [0, \infty)$.

(1) $\inf_{x\in\Omega} N_2(x, 0) > 0$,
(2) $\inf_{x\in\Omega} u(x, 0) > C^-$, $\inf_{x\in\Omega} N_1(x, 0) > 0$ and $f(s)$ is strictly monotone increasing on $[C^-, C^- + \delta_0]$.

In fact, it is clear that we can take $T_1 = 0$ for the case (1). As for the **Case (2)**, by assumption, there exist $t_1 > 0$ and $\bar{\rho}_1 > 0$ such that

$$f(u(x, t)) N_1(x, t) \geq \bar{\rho}_1 > 0 \quad \text{for all } t \in [0, t_1].$$

Here we note that N_2 satisfies

$$\partial_t N_2(x, t) \geq f(u(x, t)) N_1(x, t) - g^* N_2(x, t).$$

Hence we get

$$N_2(x, t) \geq e^{-g^*t} \int_0^t e^{g^*s} f(u(x, s)) N_1(x, s) \, ds \geq \bar{\rho}_1 t \quad \text{for all } t \in (0, t_1].$$

Thus we can take $T_1 = t_1$ and $\rho_1 = \bar{\rho}_1 t_1$.

Complete Swelling

We here consider the case where $C^- < \alpha$.

Theorem 5.4 *Let $C^- < \alpha$, then there exists some $T_c > 0$ such that the following exponential convergences hold.*

$$N_1(x, t) \xrightarrow{t \to \infty} 0 \qquad\qquad in \; \mathcal{O}(e^{-\eta_1 t}) \quad for \; a.e. \; x \in \Omega,$$

$$N_2(x, t) \xrightarrow{t \to \infty} 0 \qquad\qquad in \; \mathcal{O}(e^{-\eta_2 t}) \quad for \; a.e. \; x \in \Omega,$$

$$N_3(x, t) \xrightarrow{t \to \infty} \overline{N}(x) \qquad\qquad in \; \mathcal{O}(e^{-\eta_2 t}) \quad for \; a.e. \; x \in \Omega,$$

$$\|u(t) - \alpha\|_{L^2(\Omega)}^2 \xrightarrow{t \to \infty} 0 \qquad\qquad in \; \mathcal{O}(e^{-\gamma_2 t}),$$

$$\|\nabla_x u(t)\|_{L^2(\Omega)}^2 \xrightarrow{t \to \infty} 0 \qquad\qquad in \; \mathcal{O}(e^{-\gamma_2 t}),$$

where η_1 is a positive number depending on f and $C^- + \delta_0$, η_2 is any number satisfying $0 < \eta_2 < \min(\eta_1, g(\underline{\rho}))$ with $\underline{\rho} = \underline{\rho}(T_c) > 0$, and γ_2 is any number satisfying $0 < \gamma_2 < \min(\frac{d_1}{C_F}, 2\eta_2)$. Here the terminology $v(t) \xrightarrow{t \to \infty} v^\infty$ in $\mathcal{O}(e^{-kt})$ means that there exists some constant $C > 0$ such that

$$|v(t) - v^\infty| \leqslant C e^{-kt} \quad for \; all \; t \geqslant T_c.$$

Proof The uniform convergence of $u(x, t)$ to α implies that for any $\beta \in (0, \alpha - C^-)$, there exists a finite time $T_c = T_c(\beta) > 0$ such that

$$u(x, t) \geqslant C^- + \beta > C^- \quad \forall x \in \Omega \quad \forall t \geqslant T_c.$$

In order to derive the strict positivity of $f(u(x, t))$, we have to distinguish between two cases in accordance with Condition 5.3:

(i) *Case where $\beta \geqslant \delta_0$ (δ_0 is the parameter appearing in Condition 5.3)*
     Since $u(x, t) \geqslant C^- + \beta \geqslant C^- + \delta_0$, it follows from *(iii)* of Condition 5.3 that

$$f(u(x, t)) \geqslant f(C^- + \delta_0) > 0 \quad \forall x \in \Omega \quad \forall t \geqslant T_c.$$

(ii) *Case where $0 < \beta \leqslant \delta_0$*
     For this case, since $u(x, t) \in [C^- + \beta, C^- + \delta_0]$, from *(ii)* of Condition 5.3, we get

$$f(u(x, t)) \geqslant \eta_0 := \min\{f(s); s \in [C^- + \beta, C^- + \delta_0]\} > 0 \quad \forall x \in \Omega \quad \forall t \geqslant T_c.$$

In summary we conclude

$$f(u(x,t)) \geq \eta_1 := \min \left(f(C^- + \delta_0), \eta_0 \right) > 0 \quad \forall x \in \Omega \quad \forall t \geq T_c. \tag{5.49}$$

Substituting (5.49) into (5.2), we obtain the exponential decay estimate for $N_1(x,t)$:

$$N_1(x,t) \leq N_1(x,T_c) e^{-\eta_1(t-T_c)} \leq \|\overline{N}\|_{L^\infty(\Omega)} e^{\eta_1 T_c} e^{-\eta_1 t} \quad \forall x \in \Omega \quad \forall t \geq T_c.$$

Furthermore substituting this into (5.3) and recalling (5.17), we get

$$\partial_t N_2(x,t) \leq C_0 e^{-\eta_1 t} - g(\underline{\rho}) N_2(x,t), \quad \underline{\rho} = \underline{\rho}(T_c), \quad C_0 = f^* \|\overline{N}\|_{L^\infty(\Omega)} e^{\eta_1 T_c}. \tag{5.50}$$

Let η_2 be any number satisfying $0 < \eta_2 < \min(\eta_1, g(\underline{\rho}))$. Then by (5.50), we easily get

$$\partial_t \left(e^{\eta_2 t} N_2(x,t) \right) \leq C_0 e^{-(\eta_1 - \eta_2)t} \quad \forall x \in \Omega \quad \forall t \geq T_c. \tag{5.51}$$

Hence integrating (5.51) over (T_c, t), we obtain the exponential decay of $N_2(x,t)$:

$$N_2(x,t) \leq \left(e^{\eta_2 T_c} \|\overline{N}\|_{L^\infty(\Omega)} + \frac{C_0}{\eta_1 - \eta_2} e^{-(\eta_1 - \eta_2)T_c} \right) e^{-\eta_2 t} \quad \forall x \in \Omega \quad \forall t \geq T_c.$$

In analogy to the previous case, the conservation law together with the exponential decay obtained above implies

$$N_3^\infty(x) = \overline{N}(x) \quad \text{and} \quad N_3(x,t) \xrightarrow{t \to \infty} \overline{N}(x) \quad \text{in} \quad \mathcal{O}(e^{-\eta_2 t}) \quad \forall x \in \Omega \quad \forall t \geq T_c.$$

The exponential convergence of $\|u(t) - \alpha\|^2_{L^2(\Omega)}$ and $\|\nabla_x u(t)\|^2_{L^2(\Omega)}$ can be derived from the same reasoning as for the partial swelling case in Theorem 5.2 with $g(\underline{\rho})$ replaced by η_2.

Remark 5.8 If $C^- < \alpha$ and $u_0(x) \geq \alpha$, then assertions of Theorem 5.4 hold true without assuming any structure conditions on f and g except Condition 5.1. In fact, let $\underline{u}^\alpha(x,t) \equiv \alpha$, then $\underline{u}^\alpha$ satisfies

$$\partial_t \underline{u}^\alpha(x,t) - d_1 \Delta_x \underline{u}^\alpha = 0 \leq d_2 g(\underline{u}^\alpha) N_2 \quad \text{in } \Omega, \quad -\partial_\nu \underline{u}^\alpha = a(x)(\underline{u}^\alpha - \alpha) \quad \text{on } \partial\Omega.$$

Then from Proposition 5.1, we derive that $u(x,t) \geq \underline{u}^\alpha(x,t) \equiv \alpha$ for $a.e. x \in \Omega$ and all $t > 0$. Hence we can repeat the same arguments in the proof of Theorem 5.4.

5.5 Numerical Illustrations

We illustrate the previous results on longtime behavior with numerical simulations. For this, we have to specify appropriate functions $f(u)$ and $g(u)$. Following [5, 7] we choose

$$f(u) = \begin{cases} 0, & 0 \leqslant u \leqslant C^-, \\ \frac{f^*}{2}\left(1 - \cos \frac{(u-C^-)\pi}{C^+-C^-}\right), & C^- \leqslant u \leqslant C^+, \\ f^*, & u > C^+, \end{cases}$$

and

$$g(u) = \begin{cases} \frac{g^*}{2}\left(1 - \cos \frac{u\pi}{C^+}\right), & 0 \leqslant u \leqslant C^+, \\ g^*, & u > C^+. \end{cases}$$

The model parameters used are summarized in Table 5.1. They have been chosen primarily to support, demonstrate, and emphasize the mathematical results. As in [5], as domain $\Omega \subset \mathbb{R}^2$ we choose a disc with diameter 1. In our simulations we use the radially symmetric initial data

$$u(x, 0) = 2C^+\left[\left(1 - \sqrt{x_1^2 + x_2^2}\right)\left(1 + \sqrt{x_1^2 + x_2^2}\right)\right]^4, \quad r := \sqrt{x_1^2 + x_2^2} \quad x \in \Omega$$

and

$$N_1(x, 0) = 1, \quad N_2(x, 0) = 0, \quad N_3(x, 0) = 0, \quad x \in \Omega,$$

i.e., we assume that initially swelling has not yet been initiated.

According to Theorems 5.2, 5.3, and 5.4, important for the qualitative behavior of solutions is the external Ca^{2+} concentration value α, relative to the swelling induction threshold C^-. This is the parameter that we vary in our simulations. We choose $\alpha \in \{0, 10, 17, 25, 100, 250\}$. The first of these values reflects the situation in Theorem 5.3, the next two values represent the case $0 < \alpha < C^-$ of Theorem 5.2,

Table 5.1 Default parameter values, cf also [7] (with permission from AIMS)

Parameter	Symbol	Value	Remark
Lower (initiation) swelling threshold	C^-	20	Varied in some simulations
Upper (maximum) swelling threshold	C^+	200	
Maximum transition rate for $N_1 \rightarrow N_2$	f^*	1	
Maximum transition rate for $N_2 \rightarrow N_3$	g^*	1	
Diffusion coefficient	d_1	0.2	Varied in some simulations
Feedback parameter	d_2	30	

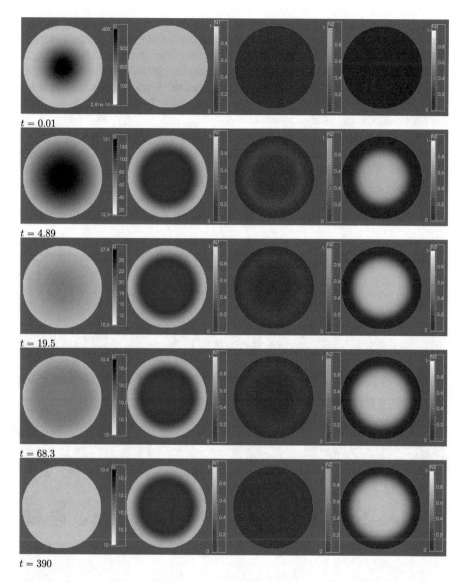

Fig. 5.4 Model simulation with $\alpha = 10 < C^-$: Shown are u, N_1, N_2, N_3 for selected times (with permission from AIMS)

and the three largest values correspond to $C^- < \alpha$ as in Theorem 5.4. Note that for the largest value we have $\alpha > C^+$, whereas $C^- < \alpha < C^+$ holds for the other two values.

In Fig. 5.4 we show for selected time instances u, N_1, N_2, N_3 for the simulation with external calcium ion concentration $\alpha = 10 < C^-$. The numerical results confirm the analysis in Theorem 5.2: The calcium ion concentration eventually attains $u = \alpha$ everywhere. The unswollen mitochondrial population N_1 remains unchanged after some initial period. In particular we note that in a layer close to the boundary almost no swelling is induced, whereas in the inner core swelling is induced everywhere. The mitochondrial population N_2 in the intermediate swelling state eventually goes to 0 everywhere. Finally, the completely swollen mitochondria attain values close to 1 in the inner core and close to 0 in an outer layer at the boundary, mirroring the distribution of N_1.

For the case $\alpha = 0$ we show in Fig. 5.5 the spatial distribution of unswollen mitochondria N_1 for selected time instances. As predicted in Theorem 5.3, this distribution does not change after some finite time. Note that the second part of the assertion of Theorem 5.3, an assertion on the mitochondria in the swelling stage, N_2, does not apply to the case of our initial conditions, which are chosen such that at the boundary of the domain $u = 0$. Since also the boundary condition enforces

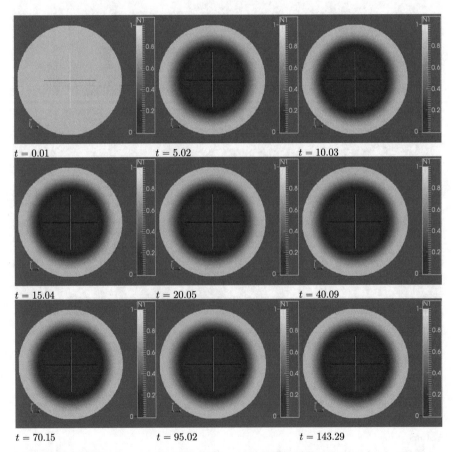

$t = 0.01$ $t = 5.02$ $t = 10.03$

$t = 15.04$ $t = 20.05$ $t = 40.09$

$t = 70.15$ $t = 95.02$ $t = 143.29$

Fig. 5.5 Model simulation with $\alpha = 10 < C^-$: Shown is N_1 for selected times (with permission from AIMS)

very small values for u at the boundary for $t > 0$, the swelling threshold C^- is never exceeded there, wherefore the hypothesis of the theorem that $N_2 > \rho_1$ almost everywhere for some positive ρ_1 is not satisfied. To illustrate this claim, we ran a second set of simulations, with different initial conditions for u, chosen such that initially $u > 0$ everywhere in the domain. More specifically we used initial data defined using

$$u_0(x) = u_{\text{base}} + 2C^- \frac{\tilde{u}(x)}{\|\tilde{u}\|_\infty}$$

where the heterogeneity $\tilde{u}$ is defined as

$$\tilde{u}(x) = 2e^{-2\sqrt{x_1^2 + x_2^2}} + \sin\left((x + y)\pi\right) + 1$$

and the base concentration $u_{\text{base}} \in \{0, 15, 50, 100, 150, 220\}$ was varied. In Fig. 5.6 we plot the minimum values of N_2 in Ω as a function of t for these different choices. We observe that in all cases N_2 is bounded from below by a constant that depends on the initial data. For the base concentrations $u_{\text{base}} > C^-$, the minimum value of

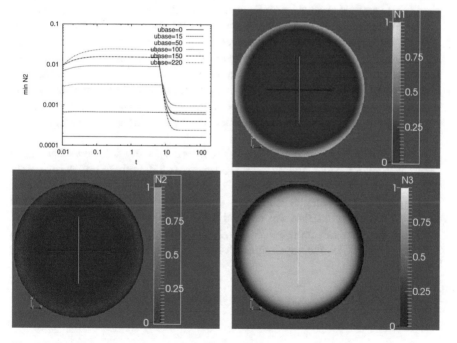

Fig. 5.6 Simulation to illustrate partial swelling in Theorem 5.3, using initial data (refT2init:eq): shown is the minimum value of N_2 as a function of time for different base calcium ion concentrations u_{base} (top left), along with the steady state distributions for N_1 (top right), N_2 (bottom left), and N_3 (bottom right) in the case $u_{\text{base}} = 100$ (with permission from AIMS)

N_2 plateaus first at some high level (the higher the higher u_{base}) and then at some T^* (the smaller the higher u_{base}) begins to drop to a lower value (the lower the higher u_{base}) that it eventually attains. Also in Fig. 5.6 we plot for $u_{base} = 100$ the spatial distribution of mitochondrial populations N_1, N_2, N_3 at steady state, showing that $N_1 > 0$ a layer close to the boundary, illustrating partial swelling. Not that the spatial heterogeneities of the initial data of u have been largely obliterated and nearly, but not exactly, radially symmetric mitochondrial populations are found. The lack of complete radial symmetry is due to the fact that close to the boundary u drops below the swelling induction threshold C^- quickly, preventing further initiation of swelling there. Due to diffusion, the calcium ion concentration eventually attains 0 everywhere (data not shown).

To verify the long-term exponential convergence of the solutions we plot in Fig. 5.7 for each of our choices of α the mitochondrial population densities N_1 and N_2 as a function of time for three different points of the domain that lay on a line through its center: point A is close to the boundary, B half way between the boundary and the center, and C is close to the center. In all cases, for large enough t these curves negative sloped lines in the logscale, indicating exponential decrease. In case $\alpha = 250$ very rapidly the calcium ion concentration is above C^+ in the entire domain leading to maximum swelling rates everywhere, whence the corresponding curves in all points overlay each other. In the cases $\alpha = 25$ and $\alpha = 100$ complete swelling occurs, i.e., both N_1 and N_2 go to 0 in all three points, where the curves of corresponding populations have the same or similar slope. In the cases $\alpha = 10$ and $\alpha = 17$, where partial swelling is observed, N_1 attains horizontal tangents for large t whereas the curves for N_2 decline. Note that convergence of N_2 is much slower for the lower external calcium ion concentration $\alpha = 10$ than for $\alpha = 17$. In the case $\alpha = 0$ eventually both N_1 and N_2 have horizontal tangents. In the two simulations with lowest external calcium ion concentrations, $\alpha = 0$ and $\alpha = 10$ no swelling is induced at the point A close to the boundary, and N_1 remains at unity there, whereas $N_2 = N_3 = 0$.

The simulation for case $\alpha > C^-$, i.e., Theorem 5.4, in Fig. 5.7 shows that for the larger calcium ion concentrations, $\alpha = 100$ and $\alpha = 250$ both N_1 and N_2 converge to 0 and thus complete swelling occurs. For the remaining case $\alpha = 25 > C^-$, i.e., the case with lowest external calcium ion concentration, N_1 and N_2 also eventually decrease, but the simulation was stopped before both the populations approached 0. Comparing the cases for $\alpha \in \{25, 100, 250\}$ shows, as suggested by the analysis that the rate at which the populations N_1 and N_2 decline depend on the boundary data.

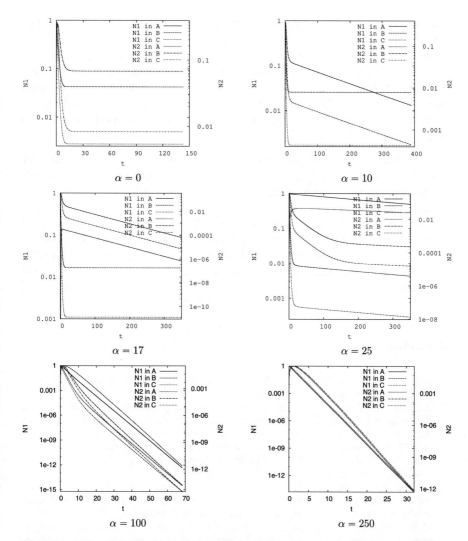

Fig. 5.7 Mitochondria populations N_1 and N_2 as a function of time in three points of the domain on a line through the center point: A (close to the boundary), B (half way between boundary and center), C (in the center), for six different values of the external calcium ion concentration α (with permission from AIMS)

Conclusion

Biologically the convergence of $u(x, t)$ to α is exactly the result we expected. Additional Ca^{2+} is removed from the cell and the calcium gradient is again stabilized. Here it is interesting to take a look at the relation of α and the swelling threshold C^-. Based on the classification of partial and complete swelling we obtained for the in vitro model, we have an analogous classification here, i.e., $\alpha > C^-$ leading to complete swelling and $\alpha < C^-$ inducing partial swelling.

However, $u(x, t) \equiv \alpha$ is also the situation when the whole system is in rest, that means when we do not have any induction of mitochondrial swelling. The condition that α being greater than C^- would hence imply that without any outer influence the cell dies immediately by apoptosis due to all mitochondria getting swollen.

Now the distinction between partial and complete swelling is only dependent on the parameter values:

- A low diffusion rate d_1 together with a high feedback rate d_2 can lead to **complete swelling**, when the initially high calcium peak is not dissolved too fast and the resulting calcium wave reaches every mitochondrion with a sufficiently high amplitude bigger than C^-.
- On the other hand, the opposite direction of high diffusion and small feedback brings forward the fast diffusion of calcium, which then leads to **partial swelling**.
- As we will see in the numerical simulations, this distinction is also highly dependent on the boundary parameter a, which determines the speed of the calcium efflux.

References

1. B. Alberts, D. Bray, K. Hopkin, A. Johnson, J. Lewis, M. Raff, K. Roberts, P. Walter, *Essential Cell Biology* (Garland Science, New York, 2013)
2. H. Brézis, Monotonicity methods in Hilbert spaces and some applications to nonlinear partial differential equations, in *Contributions to Nonlinear Functional Analysis*, pp. 101–156 (1971)
3. H. Brézis, *Opérateurs maximaux monotones et semi-groupes de contractions dans les espaces de Hilbert*, French. North-Holland Mathematics Studies. 5. Notas de matematica (50). Amsterdam-London: North-Holland Publishing Company; New York: American Elsevier Publishing Company, vol. 183, p. Dfl. 25.00; ca. $ 8.80 (1973)
4. A. Deniaud, E. Maillier, D. Poncet, G. Kroemer, C. Lemaire, C. Brenner, et al., Endoplasmic reticulum stress induces calcium-dependent permeability transition, mitochondrial outer membrane permeabilization and apoptosis. *Oncogene* **27**(3), 285–299 (2008)
5. M. Efendiev, M. Ôtani, H.J. Eberl, A coupled PDE/ODE model of mitochondrial swelling: Large-time behavior of the homogeneous Dirichlet problem. J. Coupled Syst. Multiscale Dyn. **3**(2), 122–134 (2015)
6. M. Efendiev, M. Ôtani, H.J. Eberl, Mathematical analysis of an *in vivo* model of mitochondrial swelling. English In: Discrete Contin. Dyn. Syst. **37**(7), 4131–4158 (2017)
7. S. Eisenhofer, A coupled system of ordinary and partial differential equations modeling the swelling of mitochondria. PhD thesis, Technische Universität München, 2013
8. S. Eisenhofer, F. Toókos, B.A. Hense, S. Schulz, F. Filbir, H. Zischka, A mathematical model of mitochondrial swelling. BMC Res. Not. **3**(1), 1 (2010)
9. S. Eisenhofer, M.A. Efendiev, M. Otani, S. Schulz, H. Zischka, On a ODE–PDE coupling model of the mitochondrial swelling process. Discrete Contin. Dyn. Syst. Ser. B **20**(4), 1031–1058 (2015)
10. S. Naghdi, M. Waldeck-Weiermair, I. Fertschai, M. Poteser, W.F. Graier, R. Malli, Mitochondrial Ca2+ uptake and not mitochondrial motility is required for STIM1-Orai1-dependent store-operated Ca2+ entry. J. Cell. Sci. **123**(15), 2553–2564 (2010)

11. M. Otani, Nonmonotone perturbations for nonlinear parabolic equations associated with subdifferential operators, Cauchy problems. J. Differ. Equ. **46**(2), 268–299 (1982)

12. R. Rizzuto, T. Pozzan, Microdomains of intracellular Ca2+: molecular determinants and functional consequences. Physiol. Rev. **86**(1), 369–408 (2006)

13. R. Rizzuto, S. Marchi, M. Bonora, P. Aguiari, A. Bononi, D. De Stefani, C. Giorgi, S. Leo, A. Rimessi, R. Siviero, et al., Ca 2+ transfer from the ER to mitochondria: when, how and why. Biochim. Biophys. Acta (BBA)-Bioenerg. **1787**(11), 1342–1351 (2009)

14. D. Ron, Translational control in the endoplasmic reticulum stress response. J. Clin. Invest. **110**(10), 1383–1388 (2002)

15. J. Sneyd, R. Bertram, *Tutorials in Mathematical Biosciences: Mathematical Modelling of Calcium Dynamics and Signal Transduction. II*, vol. 1867 (Springer Science and Business Media, Berlin, 2005)

16. H. Zischka, N. Larochette, F. Hoffmann, D. Hamoller, N. Jagemann, J. Lichtmannegger, L. Jennen, J. Muller-Hocker, F. Roggel, M. Gottlicher, et al., Electrophoretic analysis of the mitochondrial outer membrane rupture induced by permeability transition. Anal. Chem. **80**(13), 5051–5058 (2008)

Chapter 6
The Swelling of Mitochondria: Degenerate Diffusion

In the previous chapters mathematical modeling and analysis of mitochondria swelling in vitro, in vivo, and in silico took into account spatial effects governed by standard diffusion. A classical property of the heat equation is its *infinite speed of propagation*, i.e., for every $t > 0$ the solution $u(t)$ starting from nonnegative initial values u_0 gets automatically positive on the whole domain. We saw this fact in Proposition 4.1 and the reason becomes reliable by the divergence form representation

$$\Delta_x u = \text{div}(1 \cdot \nabla_x u)$$

with the constant diffusion coefficient $D \equiv 1$ determining the speed of diffusion. However, from the application point of view this property is not realistic, since there are no substances that can diffuse infinitely fast. Imagine a huge domain Ω and the calcium source u_0 being a delta distribution. Then biologically it is not possible that after an arbitrary small time the calcium reaches any point of the domain.

6.1 Mathematical Analysis of Mitochondria Swelling in Porous Medium

In order to obtain a more realistic setting, in the following the standard Laplacian is not the method of choice anymore. It is replaced by the porous medium operator

$$\Delta_x u^m = \text{div}\left(m u^{m-1} \nabla_x u\right) \quad \text{with } m > 1, \tag{6.1}$$

which is a nonlinear generalization of $\Delta_x u$ representing the case $m = 1$. For more general information the reader is referred to [6], where the mathematical theory of

© Springer Nature Switzerland AG 2018
M. Efendiev, *Mathematical Modeling of Mitochondrial Swelling*,
https://doi.org/10.1007/978-3-319-99100-9_6

the porous medium equation

$$\partial_t u = \Delta_x u^m$$

and its applications are studied.

The nonlinear expression (6.1) is only defined for nonnegative functions, since m does not have to be an integer. In our application u represents the calcium concentration, which is biologically restricted to be nonnegative.

Finite Propagation Speed

In contrast to the standard Laplace operator now

$$\Delta_x u^m = \text{div}(D(u)\nabla_x u), \tag{6.2}$$

where the diffusion coefficient

$$D(u) := mu^{m-1}, \quad m > 1$$

is dependent upon u and not constant anymore. That means $D(u)$ is only positive for locations where calcium is already present. This has a huge influence on the speed of propagation. As it can, e.g., be read in [6], starting from a compactly supported initial value u_0, the solution $u(t)$ has compact support for every $t > 0$, which signifies *finite speed of propagation*. Furthermore the sets

$$B(t) := \{x \in \Omega : u(x,t) > 0\} \subset \Omega$$

satisfy

$$\bigcup_{t>0} B(t) = \Omega \quad \text{and} \quad B(t_2) \subset B(t_1) \quad \text{for } t_2 > t_1.$$

The diffusion coefficient vanishes outside the set $B(t)$ if $m > 2$ and therefore the rate of diffusion is extremely small near the boundary of $B(t)$, hence the set expands slowly. On the other hand, if $m = 2$ representing the Laplacian, the diffusion coefficient remains identically constant in Ω; therefore in this case supp $u(\cdot, t)$ spreads all over $\overline{\Omega}$ instantly.

Figure 6.1 depicts the differences between infinite and finite propagation speed and makes clear why the degenerate diffusion (6.2) is more suitable to represent the process in reality compared to the standard Laplacian with infinite speed.

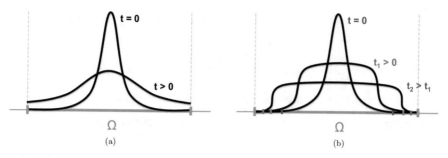

Fig. 6.1 Propagation of compactly supported initial data. (**a**) Infinite speed. (**b**) Finite speed

Degeneracy

The most important property of this new type of equation is its degeneracy, i.e., the diffusion coefficient depends on u in a sense that it vanishes for $u = 0$. In that case the PDE (6.2) is not parabolic anymore and we name it a *degenerate parabolic equation*. Understandably the degeneracy is challenging for the mathematical analysis and in the following we will see that we need additional theoretical concepts to treat this type of equation.

Thus, we analyze the swelling of mitochondria on a bounded domain $\Omega \subset \mathbb{R}^n$ and consider the following coupled PDE-ODE system determined by a constant $m > 1$ and nonnegative model functions f and g:

$$\partial_t u \quad = \quad d_1 \Delta_x u^m + d_2 g(u) N_2 \tag{6.3}$$

$$\partial_t N_1 \quad = \quad -f(u) N_1 \tag{6.4}$$

$$\partial_t N_2 \quad = \quad f(u) N_1 - g(u) N_2 \tag{6.5}$$

$$\partial_t N_3 \quad = \quad g(u) N_2 \tag{6.6}$$

with diffusion constant $d_1 > 0$ and feedback parameter $d_2 > 0$. The initial condition $u(x, 0)$ describes the added amount of Ca^{2+} to induce the swelling process.

Boundary condition:

$$u|_{\partial\Omega} = 0 \quad \text{a.e. } t \in (0, T), \tag{6.7}$$

where the bounded domain Ω can be either a test tube or the whole cell. Note that by virtue of (6.4)–(6.6), the total mitochondrial population

$$\overline{N}(x, t) := N_1(x, t) + N_2(x, t) + N_3(x, t)$$

does not change in time, that is, $\partial_t \bar{N}(x, t) = 0$, and is given by the sum of the initial data:

$$\overline{N}(x, t) = \overline{N}(x) := N_{1,0}(x) + N_{2,0}(x) + N_{3,0}(x) \quad \forall t \geqslant 0 \ \forall x \in \Omega. \tag{6.8}$$

We now give precise mathematical assumptions on f and g.

Condition 6.1 The model functions $f : \mathbb{R} \to \mathbb{R}$ and $g : \mathbb{R} \to \mathbb{R}$ have the following properties:

(i) Nonnegativity:

$$f(s) \geqslant 0 \qquad \forall s \in \mathbb{R},$$

$$g(s) \geqslant 0 \qquad \forall s \in \mathbb{R}.$$

(ii) Boundedness:

$$f(s) \leqslant f^* < \infty \qquad \forall s \in \mathbb{R},$$

$$g(s) \leqslant g^* < \infty \qquad \forall s \in \mathbb{R} \qquad \text{with } f^*, g^* > 0.$$

(iii) Lipschitz continuity:

$$|f(s_1) - f(s_2)| \leqslant L_f |s_1 - s_2| \qquad \forall s_1, s_2 \in \mathbb{R},$$

$$|g(s_1) - g(s_2)| \leqslant L_g |s_1 - s_2| \qquad \forall s_1, s_2 \in \mathbb{R} \qquad \text{with } L_f, L_g \geqslant 0.$$

Well Posedness and Asymptotic Behavior of Solutions

Theorem 6.1 *Let $\Omega \subset \mathbb{R}^n$ be bounded. Assume Condition (6.1), then it holds:*

1. *For all initial data $u_0 \in H^{-1}(\Omega)$ and $N_{i,0} \in L^\infty(\Omega)$ ($i = 1, 2, 3$), the system (6.3)–(6.6) with boundary condition (6.7) possesses a unique global solution (u, N_1, N_2, N_3) such that*

$$u \in C([0, T]; H^{-1}(\Omega)) \cap L^{m+1}(0, T; L^{m+1}(\Omega)),$$

$$\sqrt{t}\,\partial_t u, \sqrt{t}\,\Delta_x u \in L^2([0, T]; H^{-1}(\Omega)),$$

$$N_i \in L^\infty([0, T]; L^\infty(\Omega)) \ (i = 1, 2, 3) \quad \text{for all } T > 0.$$

2. *Assume further that u_0, $N_{1,0}$, $N_{2,0}$, and $N_{3,0} \geqslant 0$. Then the solution (u, N_1, N_2, N_3) preserves nonnegativity. Furthermore, N_1, N_2, and N_3 are uniformly bounded in $\Omega \times [0, \infty)$.*

3. We have the strong convergence results:

$$N_1(t) \xrightarrow{t \to \infty} N_1^\infty \geq 0 \qquad\qquad \text{in } L^p(\Omega), \qquad 1 \leq p < \infty, \qquad (6.9)$$

$$N_2(t) \xrightarrow{t \to \infty} N_2^\infty \geq 0 \qquad\qquad \text{in } L^p(\Omega), \qquad 1 \leq p < \infty, \qquad (6.10)$$

$$N_3(t) \xrightarrow{t \to \infty} N_3^\infty \leq \|\overline{N}\|_{L^\infty(\Omega)} \quad \text{in } L^p(\Omega), \qquad 1 \leq p < \infty, \qquad (6.11)$$

$$u(t) \xrightarrow{t \to \infty} u^\infty \equiv 0 \qquad\qquad \text{in } L^{m+1}(\Omega). \qquad\qquad\qquad (6.12)$$

Proof   **The existence of solutions:**   We first note that by (6.8) the essential unknown functions can be taken as (u, N_1, N_2). Let $X_T := L^1([0, T]; L^1(\Omega))$ and define the mapping

$$\mathcal{B} : u \in X_T \mapsto N^u := \left(N_1^u, N_2^u\right) \mapsto \hat{u} = \mathcal{B}(u).$$

Here for a given $u \in X_T$, $N^u = (N_1^u, N_2^u)$ denotes the solution of the ODE problem:

$$\partial_t N^u = \left(-f(u)N_1^u, \ f(u)N_1^u - g(u)N_2^u\right) =: F^u(N^u),$$

$$N^u(x, 0) = \left(N_{1,0}(x), N_{2,0}(x)\right) \qquad\qquad\qquad (6.13)$$

and $\hat{u}$ denotes the solution of the PDE problem:

$$\partial_t \hat{u} = d_1 \Delta_x \hat{u}^m + d_2 g(\hat{u}) N_2^u, \quad \hat{u}|_{\partial\Omega} = 0, \quad \hat{u}(x, 0) = u_0(x), \qquad (6.14)$$

Since, by Condition 6.1 F^u is Lipschitz continuous from $Y = L^\infty(\Omega) \times L^\infty(\Omega)$ into itself, the Picard-Lindelöf theorem assures the existence of the unique global solution $N^u \in C([0, \infty); Y)$ of (6.13) for each $u \in X_T$. In order to show the well posedness of (6.14), we here prepare some abstract setting. Let $H = H^{-1}(\Omega)$ and define the inner product of H by

$$\|u\|_{H_0^1(\Omega)} := \|\nabla_x u\|_{L^2(\Omega)}, \quad (u, v)_H := {}_{H^{-1}(\Omega)}\!<u, \Lambda v>_{H_0^1(\Omega)}, \quad \Lambda = (-\Delta_x)^{-1}.$$

Then Λ is nothing but the duality mapping from $H^{-1}(\Omega)$ onto $H_0^1(\Omega)$. Furthermore, we put

$$\varphi(u) = \begin{cases} \frac{1}{m+1} \int_\Omega |u(x)|^{m+1} dx & \text{if } u \in L^{m+1}(\Omega), \\[2mm] +\infty & \text{if } u \in H^{-1}(\Omega) \backslash L^{m+1}(\Omega). \end{cases}$$

Then φ becomes a lower semi-continuous convex function from H into $[0, +\infty]$ and its subdifferential $\partial\varphi$ is given as follows (cf. [2] and [1]).

$$\partial\varphi(u) = -\Delta_x |u|^{m-1} u,$$

$$D(\partial\varphi) = \{u \in L^{m+1}(\Omega); \Delta_x |u|^{m-1} u \in H^{-1}(\Omega)\}$$

$$= \{u \in L^{m+1}(\Omega); \int_\Omega |u|^{m-1} |\nabla_x u|^2 dx < +\infty\}.$$

Then (6.14) is reduced to the abstract problem in $H = H^{-1}(\Omega)$:

$$\frac{d}{dt}\hat{u}(t) + d_1 \partial\varphi(\hat{u}(t)) + B(\hat{u}(t)) = 0, \quad \hat{u}(0) = u_0, \quad \text{with } B(u)(\cdot, t) = -d_2 g(u(\cdot, t)).$$
$$(6.15)$$

To assure the solvability of (6.15), we rely on an abstract result in [5]. To do this, we need to check conditions (A.1), (A.2), (A.5), and (A.6) required in [5]. It is easy to see that (A.1) is satisfied, since the embedding $H^{-1}(\Omega) \subset L^{m+1}(\Omega)$ is compact. The demiclosedness condition (A.2) of B is derived from the continuity of g. The boundedness condition (A.5) is assured, since we get

$$\|B(\hat{u})\|_{H^1(\Omega)}^2 \leqslant C \|B(\hat{u})\|_{L^2(\Omega)}^2 \leqslant C d_2^2 (g^*)^2 \|N_2^u\|_{L^\infty(\Omega)}^2 |\Omega|^2 < +\infty,$$

where C denotes a general constant independent of $\hat{u}$. Moreover we have

$$(-\partial\varphi(\hat{u}) - B(\hat{u}), \hat{u})_H + (m+1)\varphi(\hat{u}) = -(B(\hat{u}), \hat{u})_H$$

$$- (B(\hat{u}), \hat{u})_H \leqslant C \|B(\hat{u})\|_{L^2(\Omega)} \|\hat{u}\|_{L^2(\Omega)} \leqslant m\varphi(\hat{u}) + C,$$

whence follows $(-\partial\varphi(\hat{u}) - B(\hat{u}), \hat{u})_H + \varphi(\hat{u}) \leqslant C$, which assures (A.6). Thus we can apply Corollary IV in [5] which assures the existence of a global solution of (6.14) satisfying

$$\hat{u} \in C([0, T]; H^{-1}(\Omega)) \cap L^{m+1}(0, T; L^{m+1}(\Omega)),$$

$$\sqrt{t}\,\partial_t \hat{u}, \sqrt{t}\,\Delta_x \hat{u} \in L^2([0, T]; H^{-1}(\Omega)).$$

We are going to verify the uniqueness of solution. Let $\hat{u}_1, \hat{u}_2$ be solutions of (6.14) with $\hat{u}_1(0) = \hat{u}_2(0) = u_0$, then $\delta\hat{u} = \hat{u}_1 - \hat{u}_2$ satisfies

$$\partial_t \delta\hat{u}(t) = d_1 \Delta_x (|\hat{u}_1|^{m-1}\hat{u}_1 - |\hat{u}_2|^{m-1}\hat{u}_2) + d_2 (g(\hat{u}_1) - g(\hat{u}_2)) N_2^u, \quad \delta\hat{u}(0) = 0.$$
$$(6.16)$$

Then multiplying (6.16) by sign $\delta\hat{u}(t)$, we obtain by the standard argument (see, e.g., [6])

$$\frac{d}{dt}\|\delta\hat{u}(t)\|_{L^1(\Omega)} \leqslant d_2 L_g \int_\Omega |\delta\hat{u}|\,|N_2^u|\,dx \leqslant C\|\delta\hat{u}(t)\|_{L^1(\Omega)},$$

where sign $w = 1$ if $w \geqslant 0$ and sign $w = -1$ if $w < 0$. Then Gronwall's inequality yields $|\delta\hat{u}(t)|_{L^1(\Omega)} \leqslant |\delta\hat{u}(0)|_{L^1(\Omega)}\,e^{Ct} \equiv 0$, whence follows the uniqueness.

Contraction Property of $\mathcal{B}$ Here we show that $\mathcal{B}$ becomes a contraction from $L^1(0, T; L^1(\Omega))$ into itself for a sufficiently small $T > 0$. Let $\delta u = u_1 - u_2$, $\delta N_i = N_i^{u_1} - N_i^{u_2}$ (i=1,2), $\delta\hat{u} = \hat{u}_1 - \hat{u}_2$. Then we have

$$\partial_t\delta N_1 = -f(u_1)\delta N_1 - (f(u_1) - f(u_2))N_1^{u_2}, \tag{6.17}$$

$$\partial_t\delta N_2 = f(u_1)\delta N_1 + (f(u_1) - f(u_2))N_1^{u_2} - g(u_1)\delta N_2 - (g(u_1) - g(u_2))N_2^{u_2}, \tag{6.18}$$

$$\partial_t\delta\hat{u} = d_1\Delta_x(|\hat{u}_1|^{m-1}\hat{u}_1 - |\hat{u}_2|^{m-1}\hat{u}_2) + d_2 g(\hat{u}_1)\delta N_2 + d_2(g(\hat{u}_1) - g(\hat{u}_2))N_2^{u_2}. \tag{6.19}$$

Multiplying (6.17) by sign $\delta N_1(t)$, we get

$$\frac{d}{dt}\|\delta N_1(t)\|_{L^1(\Omega)} \leqslant \int_\Omega L_f |u_1(t) - u_2(t)|\,|N_1^{u_2}|\,dx \leqslant L_f \|N_1^{u_2}(t)\|_{L^\infty(\Omega)}\|\delta u(t)\|_{L^1(\Omega)}.$$

Then recalling that

$$\sup\{|N_1(t)^{u_i}| + |N_2(t)^{u_i}|;\, 0 \leqslant t \leqslant T,\, i = 1, 2\} \leqslant C_T < +\infty,$$

we have

$$\sup_{0\leqslant t\leqslant T}\|\delta N_1(t)\|_{L^1(\Omega)} \leqslant L_f C_T \int_0^T \|\delta u(t)\|_{L^1(\Omega)}\,dt. \tag{6.20}$$

Next the multiplication of (6.18) by sign $\delta N_2(t)$ gives

$$\frac{d}{dt}\|\delta N_2(t)\|_{L^1(\Omega)} \leqslant \int_\Omega f^* |\delta N_1(t)|\,dx + \int_\Omega L_f |\delta u(t)|\,|N_1^{u_2}(t)|\,dx$$
$$+ \int_\Omega L_g |\delta u(t)|\,|N_2^{u_2}(t)|\,dx.$$

Hence by virtue of (6.20), we obtain

$$
\sup_{0 \leqslant t \leqslant T} \|\delta N_2(t)\|_{L^1(\Omega)} \leqslant \underbrace{\left(T f^* C_T + L_f C_T + L_G C_T \right)}_{=: C_1(T)} \int_0^T \|\delta u(t)\|_{L^1(\Omega)} \, dt.
$$

$$(6.21)$$

Finally we multiply (6.19) by sign $\delta \hat{u}(t)$ to get

$$
\frac{d}{dt} \|\delta \hat{u}(t)\|_{L^1(\Omega)} \leqslant d_2 g^* \, |\delta N_2(t)|_{L^1(\Omega)} + d_2 L_g \, C_T \, \|\delta \hat{u}(t)\|_{L^1(\Omega)}. \qquad (6.22)
$$

Thus combining (6.22) with (6.21), we obtain

$$
\sup_{0 \leqslant t \leqslant T_0} \|\delta \hat{u}(t)\|_{L^1(\Omega)} \leqslant d_2 g^* \, C_1(T) \, e^{d_2 L_g C_T T} \int_0^{T_0} \|\delta u(t)\|_{L^1(\Omega)} \, dt,
$$

$$
\int_0^{T_0} \|\delta \hat{u}(t)\|_{L^1(\Omega)} \, dt \leqslant \underbrace{T_0 d_2 g^* \, C_1(T) \, e^{d_2 L_g C_T T}}_{C_2(T,T_0)} \int_0^{T_0} \|\delta u(t)\|_{L^1(\Omega)} \, dt \quad \forall T_0 \in (0, T].
$$

Thus choosing T_0 such that $C_2(T, T_0) = 1/2$, we can assure the existence of a local solution on $(0, T_0)$. Since the choice of T_0 does not depend on the size of the initial data, we easily see that this local solution can be continued up to $[0, T]$.

Nonnegativity of Solutions Multiplying the corresponding equations for $N_i(x, t)$ by $N_i^-(x, t) = \max(-N_1(x, t), 0)$ and using assumptions $f(u) \geqslant 0$, $g(u) \geqslant 0$, we can deduce that $d \|N_i^-(t)\|_{L^2(\Omega)}/dt \leqslant 0$, whence follows $\|N_i^-(t)\|_{L^2(\Omega)} \leqslant \|N_i^-(0)\|_{L^2(\Omega)} = 0$, i.e., $N_i(x, t) \geqslant 0$. So u satisfies

$$
\partial_t u(t) - d_1 \Delta_x |u|^{m-1} u(t) = d_2 g(u(t)) N_2(t) \geqslant 0,
$$

whence follows $u(x, t) \geqslant 0$. Furthermore the nonnegativity of $N_i(t)$ and the conservation law (6.8) imply the uniform boundedness of N_i such that $0 \leqslant N_i(x, t) \leqslant \|\bar{N}\|_{L^\infty}$ ($i = 1, 2, 3$) for a.e. $x \in \Omega$ and all $t > 0$.

Convergence From (6.4) and the nonnegativity result it holds in the point-wise sense

$$
\partial_t N_1(x, t) = -f(u(x, t)) N_1(x, t) \leqslant 0 \quad \forall t \geqslant 0 \quad \text{a.e. } x \in \Omega.
$$

Hence the sequence is nonincreasing and bounded below by 0, whence follows the convergence

$$
N_1(x, t) \xrightarrow{t \to \infty} N_1^\infty(x) \geqslant 0 \quad \text{a.e. } x \in \Omega.
$$

Furthermore, by (6.8) we get

$$|N_1^\infty(x)| \leqslant \|\overline{N}\|_{L^\infty(\Omega)},$$

$$N_1(x,t) = |N_1(x,t)| \leqslant \|\overline{N}\|_{L^\infty(\Omega)} \quad \text{a.e. } x \in \Omega, \ t \in (0,\infty).$$

Then by virtue of the Lebesgue dominated convergence theorem, we conclude that $N_1(\cdot, t)$ converges to N_1^∞ strongly in $L^1(\Omega)$ as $t \to \infty$. Thus to deduce (6.9), it suffices to use the relation

$$\|N_1(t) - N_1^\infty\|_{L^p(\Omega)}^p \leqslant \left(\|N_1(t)\|_{L^\infty(\Omega)} + \|N_1^\infty\|_{L^\infty(\Omega)}\right)^{p-1} \|N_1(t) - N_1^\infty\|_{L^1(\Omega)}.$$

As for $N_3(x,t)$, since (6.6) implies that $N_3(x,t)$ is monotone increasing, we can repeat the same argument as above to get (6.11).

Combining (6.8) with (6.9) and (6.11), we can easily deduce (6.10).

In the following we are going to show that u converges to 0 strongly in $L^{m+1}(Om)$.

To this end, we first note that the integration of (6.6) on $(0, t)$ gives

$$0 \leqslant \int_0^t g(u(x,s)) N_2(x,s)\,dt = N_3(x,t) - N_3(x,0) \leqslant \|\overline{N}\|_{L^\infty(\Omega)} \quad \forall t > 0,$$

which implies

$$\int_0^\infty |g(u(s)) N_2(s)|_{L^2(\Omega)}^2\,dt \leqslant g^* \|\overline{N}\|_{L^\infty(\Omega)}^2.$$

We take the inner product between (6.3) and $-\Delta_x u^m$ in $H = H^{-1}(\Omega)$. Here we note

$$(\partial_t u, -\Delta_x u^m)_H = (\partial_t u, \partial\varphi(u(t)))_H = \frac{d}{dt}\varphi(u(t)),$$

$$(-\Delta_x u^m, -\Delta_x u^m)_H = |\Delta_x u^m|_H^2 = \int_\Omega -\Delta_x u^m\, u^m\,dx$$

$$= m^2 \int_\Omega |u|^{2(m-1)} |\nabla_x u|^2\,dx,$$

$$(d_2 g(u) N_2, -\Delta_x u^m)_H \leqslant C |\Delta_x u^m|_H |d_2 g(u) N_2|_{L^2(\Omega)}$$

$$\leqslant \frac{d_1 m^2}{2} |\Delta_x u^m|_H^2 + \frac{d_2^2 C^2}{2 d_1 m^2} |g(u) N_2|_{L^2(\Omega)}^2,$$

where C is the embedding constant satisfying $|u|_H \leqslant C |u|_{L^2(\Omega)}$. Hence we obtain

$$\frac{1}{m+1}\frac{d}{dt}|u(t)|_{L^{m+1}(\Omega)}^{m+1} + \frac{d_1 m^2}{2}\int_\Omega |u|^{2(m-1)}|\nabla_x u|^2 dx \leqslant \frac{d_2^2 C^2}{2d_1 m^2}|g(u) N_2|_{L^2(\Omega)}^2.$$

Here noting that $m^2 u^{2(m-1)}|\nabla_x u|^2 = |\nabla_x u^m|^2$ and $|u|_{L^{m+1}(\Omega)}^{2m} \leqslant |u|_{L^{2m}(\Omega)}^{2m}|\Omega|^{\frac{m-1}{m+1}}$, we derive

$$\frac{d}{dt}|u(t)|_{L^{m+1}(\Omega)}^{m+1} + \frac{d_1(m+1)}{2}|u(t)|_{L^{m+1}(\Omega)}^{2m} \leqslant \frac{d_2^2 C^2(m+1)}{2d_1 m^2}|g(u) N_2|_{L^2(\Omega)}^2.$$

Here in order to show (6.12), we prepare the following lemma.

Lemma 6.1   *Let $f \in L^1(t_0, \infty; [0, \infty))$ and $y \in W_{loc}^{1,1}([t_0, \infty); \mathbb{R}^1)$ such that*

$$y'(t) + \gamma y(t)^{1+\delta} \leqslant f(t) \quad \forall t \geqslant t_0, \tag{6.23}$$

where $t_0 \geqslant 0$, $\gamma > 0$, $\delta > 0$.   Then we have

$$y(t) \to 0 \quad as \ t \to \infty. \tag{6.24}$$

Proof   Recalling the proof of Lemma 4.3 of [5], we easily see that

$$\phi_a(t) := (|y(a)|^{-\delta} + \gamma \delta (t-a))^{-\frac{1}{\delta}} + \int_a^t f(s)\,ds \quad (a \in [t_0, \infty))$$

gives a super solution of (6.23) with $\phi(a) = |y(a)|$. Then by the comparison theorem, we arrive at

$$y(t) \leqslant \phi_a(t) \leqslant \hat{\phi}_a(t) := (\gamma \delta(t-a))^{-\frac{1}{\delta}} + \int_a^\infty f(s)\,ds \quad \forall t \in (a, \infty), \ \forall a \in [t_0, \infty).$$

Hence by taking $a = t/2$ and letting $t \to \infty$, we can derive (6.24).

Thus applying Lemma 6.1 with $y(t) = |u(t)|_{L^{m+1}(\Omega)}^{m+1}$, $t_0 = 0$, $\gamma = \frac{d_1(m+1)}{2}$, $\delta = \frac{m-1}{m+1} > 0$, we deduce (6.12).

In order to analyze more minute asymptotic behavior of solutions we need the uniform convergence of u. To this end, we prove Propositions 6.1–6.2, see below.

Let us consider the following equation for the calcium concentration u:

$$\begin{cases} \partial_t u = d_1 \Delta_x u^m + d_2 g(u) N_2(x, t), & x \in \Omega, \\ u|_{\partial\Omega} = 0, & m > 1, \end{cases} \tag{6.25}$$

where a given function g satisfies Condition 6.1.

Proposition 6.1 *Let Ω^* be a bounded domain such that $\Omega \Subset \Omega^*$ and consider the following BVP:*

$$\begin{cases} -\Delta_x \varphi(x) = \varphi^{\frac{1}{m}}(x), & x \in \Omega^*, \quad m > 1, \\ \varphi|_{\partial \Omega^*} = 0. \end{cases} \tag{6.26}$$

Then (6.26) admits a unique solution with $\varphi \in W^{2,p}(\Omega^*)$ for all $p > 1$ (hence in $\varphi \in C(\overline{\Omega})$) such that $\varphi(x) > 0$ for all $x \in \Omega^*$, which implies that there exists $\mu > 0$ such that

$$\varphi(x) \geqslant \mu > 0 \quad \text{for all } x \in \overline{\Omega}.$$

Proof Set

$$J(\varphi) := \frac{1}{2} \int_{\Omega} |\nabla_x \varphi(x)|^2 \, dx - \frac{m}{m+1} \int_{\Omega} |\varphi(x)|^{\frac{m+1}{m}} \, dx, \quad m > 1.$$

Since $m > 1$ we have $\frac{m+1}{m} \in (1, 2)$, hence $J(u)$ is bounded from below in $H_0^1(\Omega)$. Furthermore, by the compact embedding $H_0^1(\Omega) \subset L^{\frac{m+1}{m}}(\Omega)$, it is easy to show that functional J has a global minimizer $\varphi \in H_0^1(\Omega)$, which turns out to be unique. Since $J(\varphi) = J(|\varphi|)$, without loss of generality we can assume that $\varphi \geqslant 0$. Now, φ is a solution to (6.26). From the strong maximum principle it follows that $\varphi > 0$ in Ω^*. Since $\Omega \Subset \Omega^*$ and φ is continuous in $\overline{\Omega}$ we conclude that there exists a $\mu > 0$ such that $\varphi \geqslant \mu > 0$ in $\overline{\Omega}$. This proves Proposition 6.1.

Next we prove an L^∞ estimate for solutions to (6.25). To this end, we construct a super-solution $\overline{u}(x)$ for (6.26) using Proposition 6.1.

Proposition 6.2 *Let u be a solution to (6.25). Then there exists a constant $C > 0$ such that*

$$\sup_{t \geqslant 1} \|u(t)\|_{L^\infty(\Omega)} \leqslant C.$$

Proof The proof is based on a construction of a time-independent super-solution $\overline{u}$ for (6.25). We set $\overline{u} := C_* \left(\frac{\varphi}{\mu} \right)^{\frac{1}{m}}$ where φ and μ are from Proposition 6.1 and constant C_* will be fixed later. It is not difficult to see that $\overline{u}$ satisfies

$$\partial_t \overline{u} = 0, \quad -\Delta_x \overline{u}^m = -\frac{C_*^m}{\mu} \Delta_x \varphi = \frac{C_*^m}{\mu} \varphi^{\frac{1}{m}},$$

and

$$-d_2 g(\overline{u}) N_2 \geqslant -d_2 L_g \overline{u} \|N_2\|_{L^\infty(\Omega)} = -d_2 L_g \|N_2\|_{L^\infty(\Omega)} \mu^{-\frac{1}{m}} \varphi^{\frac{1}{m}}.$$

Hence we obtain that

$$\partial_t \overline{u} - \Delta_x \overline{u}^m - d_2 g(\overline{u}) N_2 \geq \varphi^{\frac{1}{m}} \left(\frac{C_*^m}{\mu} - d_2 L_g \|N_2\|_{L^\infty(\Omega)} \mu^{-\frac{1}{m}} C_* \right).$$

For sufficiently large $C_* \gg 1$ we obtain that

$$\partial_t \overline{u} - \Delta_x \overline{u}^m - d_2 g(\overline{u}) N_2 \geq 0.$$

In other words, $\overline{u}$ gives a super solution for (6.25). Moreover, since there exists a $t_0 \in (0, 1)$ such that (due to $H^2(\Omega) \subset L^\infty(\Omega)$)

$$\|u(t_0)\|_{L^\infty(\Omega)} \leq C_0 \leq C_0 \left(\frac{\varphi}{\mu} \right)^{\frac{1}{m}}.$$

Thus we get

$$u(x, t) \leq C_0 \left(\frac{\varphi}{\mu} \right)^{\frac{1}{m}} \quad \text{for all } t \geq t_0 \text{ and } x \in \Omega$$

and, consequently,

$$\sup_{t \geq 1} \|u(t)\|_{L^\infty(\Omega)} \leq C_0.$$

This proves Proposition 6.2.   □

Proposition 6.3 *Let all assumptions in Theorem 6.1 be satisfied. Then we have*

$$\max_{x \in \overline{\Omega}} |u(x, t)| \to 0 \quad as \ t \to \infty. \tag{6.27}$$

Proof First we note that due to Proposition 6.2 u satisfies

$$\partial_t u(x, t) = d_1 \Delta_x u^m(x, t) + f(x, t),$$
$$f(x, t) := d_2 g(u(x, t)) N_2(x, t) \in L^\infty(0, \infty; L^\infty(\Omega)).$$

Hence by virtue of a regularity result from [4], we find that there exists $\alpha \in (0, 1)$, $T_* > 0$ and C_{T_*} such that

$$\sup_{T_* \leq t < \infty} |u(t)|_{C^\alpha(\Omega)} \leq C_{T_*}. \tag{6.28}$$

Suppose here that (6.27) does not hold. Then there exists $\rho > 0$, $x_k \in \overline{\Omega}$ and $t_k \to \infty$ as $k \to \infty$ such that

$$|u(x_k, t_k)| \geq \rho > 0 \quad \forall k. \tag{6.29}$$

Then by virtue of (6.28), there exists a subsequence $\{t_{k_j}\}$ of $\{t_k\}$ such that

$$u(t_{k_j}) \to u_\infty \quad \text{strongly in } C(\overline{\Omega}) \quad \text{as } j \to \infty.$$

Here $u_\infty \equiv 0$ follows from (6.12), which contradicts (6.29).

Remark 6.1 To derive the uniform convergence of u for the nondegenerate cases, we need some additional structure conditions on f and g. However, for the degenerate case, we can derive the uniform convergence of u without these structure conditions.

Corollary 6.1 *Let the assumptions of Theorem 6.1 hold. Then partial swelling scenario occurs.*

Indeed, by the uniform convergence of $u(x, t)$ to $u^\infty \equiv 0 < C^-$ follows the existence of time $T_\rho > 0$ such that

$$u(x, t) \leqslant C^- \quad \forall x \in \Omega, \, t \geqslant T_\rho$$

and consequently

$$f(u(x, t)) \quad \forall x \in \Omega, \, t \geqslant T_\rho.$$

From the model equation (6.4) it follows immediately for all $x \in \Omega$ that

$$\partial_t N_1(x, t) = 0 \quad t \geqslant T_\rho$$

which implies

$$N_1(x, t) \equiv N_1(T_\rho, x).$$

However, in contrast to Neumann boundary condition we cannot prove that $N_2(x, t) \to 0$ as $t \to 0$. Indeed, the following proposition holds:

Proposition 6.4 *Assume that g satisfies in addition to Condition 6.1 the following condition:*

$(C)_g \quad \exists \rho_0 > 0, \, C_g > 0 \text{ s.t. } g \text{ is monotone increasing in } (0, \rho_0) \text{ and}$

$$g(s) \leqslant C_g \, s^{m+1} \quad \forall s \in (0, \rho_0].$$

Suppose that there exists $T_1 \in [0, \infty)$ and $\rho_1 > 0$ such that $N_2(x, T_1) \geqslant \rho_1$ for a.e. $x \in \Omega$. Then there exists $\rho_2 > 0$ such that

$$N_2^\infty(x) \geqslant \rho_2 \quad a.e. \, x \in \Omega.$$

We leave the details of the proof of Proposition 6.4 to the reader.

6.2 Numerical Simulation

Governing Equations The model is formulated in terms of the dependent variables u, N_1, N_2, N_3 as in [3]. We consider the case of degenerate diffusion of substrate u, written in conservative form

$$\frac{\partial u}{\partial t} = d\nabla_x \left(u^m \nabla_x u \right) + g(u)N_2$$

$$\frac{\partial N_1}{\partial t} = -f(u)N_1$$

$$\frac{\partial N_2}{\partial t} = f(u)N_1 - g(u)N_2$$

$$\frac{\partial N_3}{\partial t} = g(u)N_2$$

with

$$f(u) = \begin{cases} f^* & \text{for} \quad u > u^+ \\ f^* \left(\frac{1}{2} - \frac{1}{2} \cos \frac{(u-u^+)\pi}{u^+-u^-} \right) & \text{for} \quad u^+ > u > u^- \\ 0 & u < u^- \end{cases}$$

and

$$g(u) = \begin{cases} g^* & \text{for} \quad u > u^+ \\ g^* \left(\frac{1}{2} - \frac{1}{2} \cos \frac{u\pi}{u^+} \right) & \text{for} \quad u < u^+ \end{cases}$$

The model parameters for the reaction terms are adapted from [3], which unfortunately does not give physical units. The parameters used in the simulations here are summarized in Table 6.1.

The diffusion coefficient d and the degeneracy exponent m in (6.30) were varied in the simulation. In order to be physically consistent, so that the diffusion term has proper physical units $[M^3 L^2 T^{-1}]$, the substrate concentration u must be either non-dimensionalized w.r.t. some reference concentration u_∞, or the diffusion coefficient d must scale with u_∞^{-m}. Both approaches are equivalent. There seems to be no natural choice for u_∞, therefore I chose, somewhat arbitrarily $u_\infty = u^+$. Thus the

Table 6.1 Reaction parameters used in the simulations, from [3]

Parameter	Symbol	Value	Dimension
Diffusion coefficient	d		L^2T^{-1}
Degeneracy exponent	m	*Varied*	–
Swelling threshold	u^-	20	ML^{-3}
Saturation threshold	u^+	200	ML^{-3}
Maximum transition rate $N_1 \rightarrow N_2$	f^*	1	T^{-1}
Maximum transition rate $N_2 \rightarrow N_3$	g^*	0.1	T^{-1}

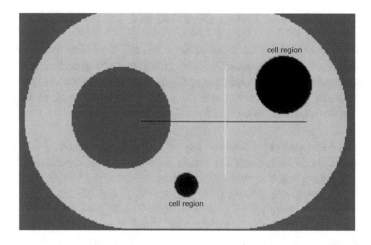

Fig. 6.2 Representation of the computational domain: The cell is oval shaped, with a nucleus in its interior. Initially increased levels of u are found in two spherical regions in the cell, which are indicated in black. The domain Ω is the oval well including the black pockets, but excluding the nucleus, i.e., the light gray and black area

diffusion coefficient reads

$$d = \frac{\delta}{(u^+)^m}$$

where δ is a given constant with units $[L^2T^{-1}]$.

Geometry The computational domain mimics the in vivo geometry used in [3]. We assume an oval shaped cell with a spherical nucleus, cf Fig. 6.2. The cell is embedded in a rectangular grid of size $1 \times 2/3$, discretized by 300×200 grid points. The model/simulation domain, in which Ω the equations are solved, is the cell without the nucleus. The boundary of Ω is divided into two parts $\partial\Omega = \Gamma_{wall} \cup \Gamma_{nucleus}$, where Γ_{wall} is the cell wall, and $\Gamma_{nucleus}$ the boundary of the nucleus.

Initial and Boundary Conditions In the simulations so far I investigated two set of boundary/initial conditions for u. For $N_{1,2,3}$ the initial conditions are always

$$N_1(x, 0) = 1, \quad N_{2,3}(x, 0) = 0, \quad x \in \Omega.$$

To specify the initial conditions we define three regions within Ω. The first two are spherical pockets of different diameters located within the cell but outside the nucleus, see Fig. 6.2. The third region is the rest of Ω without these two spherical pockets. Within each of the three regions, the initial substrate concentration is constant. We denote by u_1 the initial concentration in the larger pocket, by u_2 the initial condition in the smaller pocket, and by u_0 the initial condition in the rest of Ω.

[i] Homogeneous Neumann conditions. In the first set of simulations, homogeneous Neumann conditions are specified both on Γ_{wall} and on $\Gamma_{nucleus}$. In this case

the initial condition for u are $u_{1,2} > u^+$, $u_0 = 0$, i.e., initial data have compact support.

[ii] Nonhomogenous Dirichlet conditions at the cell wall. For the second set of simulations, nonhomogenous Dirichlet conditions are specified at Γ_{wall} and homogeneous Neumann conditions at $\Gamma_{nucleus}$. In this case we have again for the initial data $u_{1,2} > u^+$, but choose $0 < u_0 < u^-$. For the boundary concentration on Γ_{wall} we set $u|_{\Gamma_{wall}} = u_0$.

Simulation Experiments I have run several simulation of cases [i], [ii]. Included in this report are in both cases the results for a simulation experiment, in which the degeneracy exponent m was varied to take values $m = 0, 1, 2$. The first case describes nondegenerate diffusion. In both cases [i] and [ii] the initial substrate concentrations $u_{1,2}$ were varied as $u_{1,2} = ku^+$ $k = 2, 3, 4$. In case [ii], we used $u_0 = 10$ for initial and boundary conditions. Thus, in total we document here the results of 9 simulations for each case [i] and [ii]. All simulations were terminated when $t = 200$ was reached.

In Fig. 6.3 for each of the nine cases of [i] the average substrate concentrations and $N_{1,2,3}$ averaged over the domain Ω are plotted as functions of time. The corresponding results for experiment [ii] are shown in Fig. 6.4.

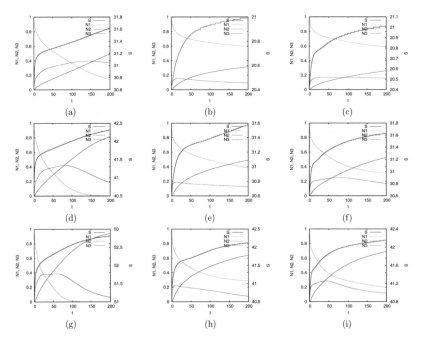

Fig. 6.3 Lumped results for the nine simulations of case [i]: the initial substrate concentration $u_{1,2}$ increases from top row to bottom row, the degeneracy exponent m from left column to right column. (**a**) $m = 0, u_{1,2} = 2$. (**b**) $m = 1, u_{1,2} = 2$. (**c**) $m = 2, u_{1,2} = 2$. (**d**) $m = 0, u_{1,2} = 3$. (**e**) $m = 1, u_{1,2} = 3$. (**f**) $m = 2, u_{1,2} = 3$. (**g**) $m = 0, u_{1,2} = 4$. (**h**) $m = 1, u_{1,2} = 4$. (**i**) $m = 2, u_{1,2} = 4$

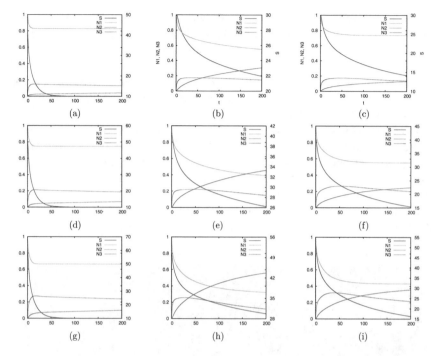

Fig. 6.4 Lumped results for the nine simulations of case [i]: the initial substrate concentration $u_{1,2}$ increases from top row to bottom row, the degeneracy exponent m from left column to right column. (**a**) $m = 0$, $u_{1,2} = 2$. (**b**) $m = 1$, $u_{1,2} = 2$. (**c**) $m = 2$, $u_{1,2} = 2$. (**d**) $m = 0$, $u_{1,2} = 3$. (**e**) $m = 1$, $u_{1,2} = 3$. (**f**) $m = 2$, $u_{1,2} = 3$. (**g**) $m = 0$, $u_{1,2} = 4$. (**h**) $m = 1$, $u_{1,2} = 4$. (**i**) $m = 2$, $u_{1,2} = 4$

In all simulation of case [i] the average substrate concentration increases with time. The average value for N_1 decreases, and the value for N_3 increases. The average value of N_2 increases first attains then a maximum and the decreases to level off. As can be expected, higher initial substrate concentrations in case [i] lead to higher substrate concentrations throughout, lower average values for N_1, and higher average values for N_3. The effect on average values for N_2 is not as clear. Note, however, that the three quantities are intimately connected since $N_2 = 1 - N_1 - N_3$. The substrate concentration at the end of the simulation period in case [i] is substantially higher for the nondegenerate case $m = 0$ than for the degenerate cases $m > 0$. In fact the average concentrations for $m = 1$ and $m = 2$ are quite similar with the one for $m = 2$ appearing slightly higher. This suggests that the initial accelerated spread in the pockets with $u > 0$ dominates. The values for N_1 at the end of the simulation are much higher still for the degenerate case than for the nondegenerate case, and N_3 is much lower. The behavior of N_2 is not as easily described in terms of m.

In case [ii] (Fig. 6.4) the Dirichlet condition with $u_0 < u^-$ induces a substrate sink at the cell wall that enforces concentrations there to be below swelling threshold. Accordingly, the average substrate concentrations there decline. The average values for N_1, N_2, and N_3 behave qualitatively as in case [i]: N_i declines, N_3 increases, N_2 increases first and then declines or levels off. Increased initial concentrations for u leads to increased values for u throughout. The average values for N_1 at the end of the simulations are larger than initial substrate concentrations. For N_2 the situation is reversed; the maximum of N_2 becomes larger as the initial substrate concentration increases. The average value for N_3 increases with increasing initial substrate concentration. In the nondegenerate case, the substrate concentration levels off quickly at bulk concentration value. In the degenerate cases this plateau is not reached within the simulation period. The average substrate concentrations for $m = 1$ are higher than for $m = 2$. At the end of the simulation N_1 is largest for the nondegenerate case $m = 0$, smallest for $m = 1$. The reverse holds for N_3. The behavior of N_2 is not as clearly described in terms of m, but it appears that the values for $m = 2$ are larger than for $m = 1$ at the end of the simulation, but for all three choices of m in a similar range.

In Fig. 6.5 we plot in detail the simulation results for case [i] simulation with $m = 1$, $u_0 = 3$, i.e., the simulation of Fig. 6.3e. Color-coded is the substrate

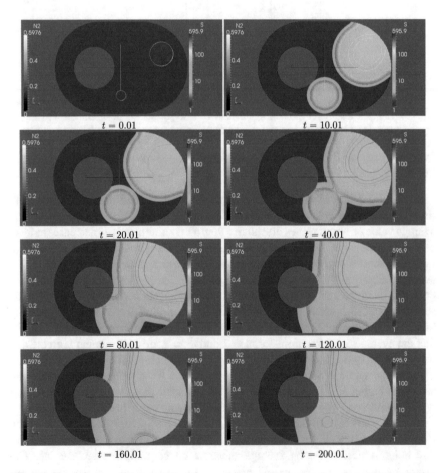

Fig. 6.5 Simulation result for case [i] with $m = 1$, $u = 3$. Shown are the substrate concentration u (colored) and the volume fraction N_2 (contour lines)

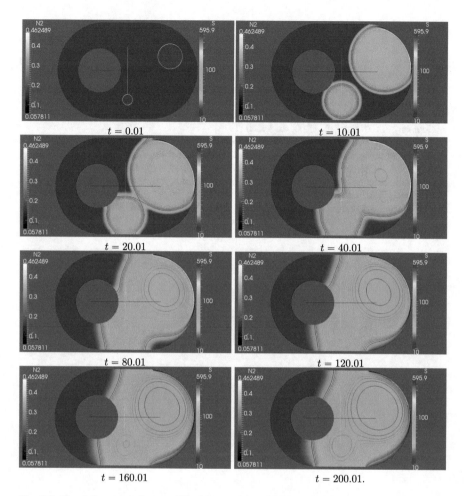

Fig. 6.6 Simulation result for case [ii] with $m = 1$, $u = 3$. Shown are the substrate concentration u (colored) and the volume fraction N_2 (contour lines)

concentration u for selected time instances. The grayscale contour lines depict N_2. Due to the porous medium degeneracy and the initial data with compact support, u propagates with a finite interface speed. Behind the interface of u a highly heterogeneous pattern of N_2 is observed.

The corresponding simulation for case [ii] is visualized in Fig. 6.6, again for $m = 1$, $u_0 = 3$, i.e., the simulation of Fig. 6.4e. Color-coded is the substrate concentration u for selected time instances. The grayscale contour lines depict N_2. In this case the initial data do not have compact support. Therefore, no sharp interface propagates but the region of increased u that propagates into Ω appears somewhat smeared out. Note that along the cell wall the concentration of u is set to 10, which induces gradients there.

References

1. H. Brézis, Monotonicity methods in Hilbert spaces and some applications to nonlinear partial differential equations. *Contributions to Nonlinear Functional Analysis*, pp. 101–156 (1971)
2. H. Brézis, *Opérateurs maximaux monotones et semi-groupes de contractions dans les espaces de Hilbert*, French. North-Holland Mathematics Studies. 5. Notas de matematica (50). Amsterdam-London: North-Holland Publishing Company; New York: American Elsevier Publishing Company, Inc. 183 p. Dfl. 25.00; ca. $ 8.80 (1973)
3. S. Eisenhofer, A coupled system of ordinary and partial differential equations modeling the swelling of mitochondria. PhD thesis, Technische Universität München, 2013
4. A. Ivanov, Estimates of the Hölder constant for generalized solutions of degenerate parabolic equations. Russian. Zap. Nauchn. Semin. Leningr. Otd. Mat. Inst. Steklova **152**, 21– 44 (1986)
5. M. Otani, Nonmonotone perturbations for nonlinear parabolic equations associated with subdifferential operators, Cauchy problems. J. Differ. Equ. **46**(2), 268–299 (1982)
6. J.L. Vázquez, *The Porous Medium Equation: Mathematical Theory* (Oxford University Press, Oxford, 2007)

Chapter 7
The Spatial Evolution of Mitochondria: PDE-PDE Systems

7.1 Well Posedness

In the previous chapters we greatly used the assumption that mitochondria do not diffuse within the cell. However, there are indications that mitochondria do move under certain circumstances, e.g., dependent of the cell cycle [4]. This spatial evolution is taken into account in this chapter which, as we will see below, leads to a PDE-PDE system. That means that mitochondrial subpopulations $N_i(x, t)$, in contrast to previous chapters, satisfy partial and not ordinary differential equations. Indeed, we consider the following PDE-PDE systems:

$$\partial_t \begin{pmatrix} u \\ N_1 \\ N_2 \\ N_3 \end{pmatrix} + \begin{pmatrix} -d_1 \Delta_x u \\ -d_2 \Delta_x N_1 \\ -d_3 \Delta_x N_2 \\ -d_4 \Delta_x N_3 \end{pmatrix} + \begin{pmatrix} -d_2 g(u) N_2 \\ f(u) N_1 \\ -f(u) N_1 + g(u) N_2 \\ -g(u) N_2 \end{pmatrix} = \mathbf{0}. \tag{7.1}$$

We denote $H := (L^2(\Omega))^4$ and

$$\mathbf{v} = \begin{pmatrix} u \\ N_1 \\ N_2 \\ N_3 \end{pmatrix}, \quad A\mathbf{v} := - \begin{pmatrix} d_1 \Delta_x u \\ d_2 \Delta_x N_1 \\ d_3 \Delta_x N_2 \\ d_4 \Delta_x N_3 \end{pmatrix}, \quad B\mathbf{v} := \begin{pmatrix} -d_2 g(u) N_2 \\ f(u) N_1 \\ -f(u) N_1 + g(u) N_2 \\ -g(u) N_2 \end{pmatrix},$$

$$\dot{D}(A) = \left\{ u \in H^2(\Omega) \mid \frac{\partial u}{\partial n}\Big|_{\partial \Omega} = 0 \right\}.$$

We impose both boundary conditions

$$\frac{\partial \mathbf{v}}{\partial n}\Big|_{\partial \Omega} = 0$$

© Springer Nature Switzerland AG 2018
M. Efendiev, *Mathematical Modeling of Mitochondrial Swelling*,
https://doi.org/10.1007/978-3-319-99100-9_7

and

$$v|_{t=0} = v_0(x). \tag{7.2}$$

Then the system (7.1)–(7.2) can be rewritten as

$$\begin{cases} \partial_t v + Av + Bv = 0 & (7.3) \\ v(x, 0) = v_0(x) & (7.4) \\ \dfrac{\partial v}{\partial n}\bigg|_{\partial\Omega} = 0. & (7.5) \end{cases}$$

For simplicity of presentation we impose analogous conditions on f and g, although depending on boundary conditions they can change as we have seen in the previous chapters. The main result of this section is the following theorem.

Theorem 7.1 *Let f and g satisfy Condition 4.1. Then for any $v_0 = (u_0, N_{1,0}, N_{2,0}, N_{3,0}) \in (L^2(\Omega))^4$ there exists unique solution of (7.3)–(7.4)–(7.5), such that*

$$\begin{cases} v \in C([0, \infty), H), \ \sqrt{t}\partial_t v, \ \sqrt{t}Av \in L^2(0, T, H), \\ \varphi(v) \in L^1(0, T), \ t\varphi(v) \in L^\infty(0, T) \ for \ any \ T > 0, \end{cases}$$

where

$$\varphi(v) := \frac{d_1}{2}\int_\Omega |\nabla_x u|^2 dx + \frac{d_2}{2}\int_\Omega |\nabla_x N_1|^2 dx$$
$$+ \frac{d_3}{2}\int_\Omega |\nabla_x N_2|^2 dx + \frac{d_4}{2}\int_\Omega |\nabla_x N_3|^2 dx.$$

Proof Note that due to Condition 4.1 we obtain

$$\|Bv\|_H^2 \leqslant d_2^2(g_*)^2\|N_2\|_{L^2(\Omega)}^2 + 2(f^*)^2\|N_1\|_{L^2(\Omega)}^2 + 2(g_*)^2\|N_3\|_{L^2(\Omega)}^2 \leqslant C\|v\|_H^2.$$

Then applying to (7.3)–(7.4)–(7.5) the result from the Chap. 6 one can prove the existence of solutions of (7.3)–(7.4)–(7.5). We omit these details and leave them to the reader.

Next we prove uniqueness. Indeed, let $v_i = (u_i, N_{1,i}, N_{2,i}, N_{3,i})$ for $i = 1, 2$ be solutions belonging to (7.1). Then

$$\partial_t(v_1 - v_2) + A(v_1 - v_2) + Bv_1 - Bv_2 = 0. \tag{7.6}$$

Multiplying both sides of (7.6) by $\delta v = v_1 - v_2$ and integrating over Ω we obtain:

$$\frac{1}{2}\frac{d}{dt}\|\delta v(t)\|_H^2 + \|A^{1/2}\delta v\|_H^2 \leq \|Bv_1 - Bv_2\|_H \cdot \|\delta v\|_H, \tag{7.7}$$

$$\|Bv_1 - Bv_2\|_H^2 \leq C\left[\|g(u_1)N_{2,1} - g(u_2)N_{2,2}\|_{L^2(\Omega)}^2\right.$$
$$\left. +\|f(u_1)N_{1,1} - f(u_2)N_{1,2}\|_{L^2(\Omega)}^2\right].$$

Note that

$$\|g(u_1)N_{2,1} - g(u_2)N_{2,2}\|_{L^2(\Omega)}^2 \leq \|g(u_1)\delta N_2\|_{L^2(\Omega)}^2 + \|g(u_1) - g(u_2)\delta N_{2,2}\|_{L^2(\Omega)}^2$$
$$\leq g_*\|\delta N_2\|_{L^2(\Omega)}^2 + L_g\|\delta u\|_H^1\|N_{2,2}\|_{H^1}. \tag{7.8}$$

Analogously, we obtain

$$\|f(u_1)N_{1,1} - f(u_2)N_{1,2}\|_{L^2(\Omega)}^2 \leq f_*\|\delta N_1\|_{L^2(\Omega)}^2 + L_f\|\delta u\|_H^1\|N_{1,2}\|_{H^1}. \tag{7.9}$$

Hence from (7.7), (7.8), and (7.9) it follows that

$$\frac{1}{2}\frac{d}{dt}\|\delta v(t)\|_H^2 + \|A^{1/2}\delta v\|_H^2 \leq C_1\|\delta v(t)\|_H^2. \tag{7.10}$$

Integrating (7.10) over $[0, T]$ and using the Gronwall inequality we have

$$\|\delta v(t)\|_H^2 \leq C_2\|\delta v(0)\|_H^2 \cdot e^{2C_2 t}. \tag{7.11}$$

Since v_1 and v_2 are the solutions of (7.3)–(7.4)–(7.5) with the same initial conditions, (7.11) leads to uniqueness of solutions of (7.3)–(7.4)–(7.5). Well posedness is proved. $\qquad\square$

Our next step is to prove "validity" of our spatial evolution mitochondria model, that is,

Proposition 7.1 *Let* $u_0(x) \geq 0$, $N_{i,0}(x) \geq 0$, $i = 1, 2, 3$. *Then any solution of* (7.3)–(7.5) *satisfies*

$$u(x, t) \geq 0, \quad N_i(x, t) \geq 0 \text{ for any } t \geq 0, \text{ a.e. } x \in \Omega, \ i = 1, 2, 3.$$

Proof Consider first $\partial_t N_1 - d_1 \Delta_x N_1 = -f(u)N_1$ and multiplying it by $N_1^-(x, t)$ and integrate over domain Ω. Then we obtain

$$-\frac{1}{2}\frac{d}{dt}\|N_1^-(t)\|_{L^2(\Omega)}^2 - d_1\|\nabla_x N_1^-(t)\|_{L^2(\Omega)}^2 = \int_\Omega f(u)|N_1^-(x, t)|^2 dx.$$

Hence

$$\frac{d}{dt}\|N_1^-(t)\|_{L^2(\Omega)}^2 + d_1\|\nabla_x N_1^-(t)\|_{L^2(\Omega)}^2 = -\int_\Omega f(u)|N_1^-(x,t)|^2 dx \leqslant 0.$$

Integrating the last inequality over $[0, t]$, $t > 0$ we have

$$\|N_1^-(t)\|_{L^2(\Omega)}^2 \leqslant \|N_1^-(0)\|_{L^2(\Omega)}^2 = 0 \implies N_1^-(x,t) \equiv 0 \text{ a.e. } x \in \Omega.$$

To prove the same property for N_2 we act in the same way, namely:

$$\partial_t N_2 + d_3 \Delta_x N_2 = f(u)N_1(x,t) - g(u)N_2(x,t).$$

Multiplying the last equality by $N_2^-(x,t)$ and integrating over Ω yield

$$-\frac{1}{2}\frac{d}{dt}\|N_2^-(t)\|_{L^2(\Omega)}^2 - d_3\|\nabla_x N_2^-(t)\|_{L^2(\Omega)}^2$$

$$= \underbrace{\int_\Omega f(u)N_1 N_2^- dx}_{\geqslant 0} + \int_\Omega g(u)|N_2^-(x,t)|^2 dx,$$

$$\frac{1}{2}\frac{d}{dt}\|N_2^-(t)\|_{L^2(\Omega)}^2 + d_3\|\nabla_x N_2^-(t)\|_{L^2(\Omega)}^2 \leqslant -\int_\Omega g(u)|N_2^-(x,t)|^2 dx \leqslant 0.$$

$$\tag{7.12}$$

Integrating (7.12) over $[0, t]$ and using the Gronwall inequality we have $N_2^-(x,t) = 0$, $\forall t > 0$ and a.e. $x \in \Omega$. Hence, $N_2(x,t) \geqslant 0$ for any $t > 0$ and a.e. $x \in \Omega$. For $N_3(x,t)$ we act in the same way. For completeness we will present a proof $N_3(x,t) \geqslant 0$ as well as $u(x,t) \geqslant 0$. Indeed, let $N_3(x,t)$ be a solution of

$$\partial_t N_3 - d_4 \Delta_x N_3 = g(u)N_2(x,t). \tag{7.13}$$

Multiplying (7.13) by $N_3^-(x,t)$ and integrating over Ω

$$-\frac{1}{2}\frac{d}{dt}\|N_3^-(t)\|_{L^2(\Omega)}^2 - d_4\|\nabla_x N_3^-(t)\|_{L^2(\Omega)}^2 = \int_\Omega g(u)N_2 N_3^- dx. \tag{7.14}$$

Since $N_2(x,t) \geqslant 0$, then from (7.14) it follows that

$$\frac{d}{dt}\|N_3^-(t)\|_{L^2(\Omega)}^2 + 2d_4\|\nabla_x N_3^-(t)\|_{L^2(\Omega)}^2 = -2\int_\Omega g(u)N_2 N_3^- dx \leqslant 0. \tag{7.15}$$

Integrating (7.15) over $[0, t]$ for any $t > 0$, we obtain

$$\|N_3^-(t)\|_{L^2(\Omega)}^2 \leqslant \|N_3^-(0)\|_{L^2(\Omega)}^2.$$

Hence $N_3^-(x, t) \equiv 0$ and as a result, $N_3(x, t) \geqslant 0$ for any $t > 0$ and a.e. $x \in \Omega$.
   Analogously multiplying

$$\partial_t u - d_1 \Delta_x u = d_2 g(u) N_2$$

by $u^-(x, t)$ and integrating over Ω and taking into account that $N_2(x, t) \geqslant 0$ for
a.e. $x \in \Omega$ and $t > 0$ we obtain

$$-\frac{1}{2}\frac{d}{dt}\|u^-(t)\|^2_{L^2(\Omega)} - d_1\|\nabla_x u^-(t)\|^2_{L^2(\Omega)} = d_2 \underbrace{\int_\Omega g(u) N_2(x, t) u^- dx}_{\geqslant 0}. \qquad (7.16)$$

and consequently integrating (7.16) over $[0, t]$ we have

$$\|u^-(t)\|_{L^2(\Omega)} \leqslant \|u^-(0)\|_{L^2(\Omega)}$$

which leads to $u(x, t) \geqslant 0$ for all $t > 0$ and a.e. $x \in \Omega$.

7.2   Asymptotic Behavior of Solutions

Having well posedness of (7.3)–(7.5) our next task is to study asymptotic behavior
of solutions as time goes to infinity. First we start asymptotics of subpopulations of
mitochondria $N_i(x, t)$. Recall that, they satisfy

$$\begin{cases} \partial_t N_1 = d_3 \Delta_x N_1 - f(u) N_1 \\ \partial_t N_2 = d_4 \Delta_x N_2 + f(u) N_1 - g(u) N_2 \\ \partial_t N_3 = d_5 \Delta_x N_2 + g(u) N_2 \end{cases} \qquad (7.17)$$

with $\frac{\partial N_i}{\partial n}|_{\partial\Omega} = 0$ and $N_i(x, 0) = N_{i,0}(x)$. Integrating over Ω of (7.17) we obtain

$$\begin{cases} \dfrac{d}{dt}\displaystyle\int_\Omega N_1(x, t)dx = -\int_\Omega f(u) N_1(x, t)dx \\ \dfrac{d}{dt}\displaystyle\int_\Omega N_2(x, t)dx = \int_\Omega f(u) N_1(x, t)dx - \int_\Omega g(u) N_2(x, t)dx \\ \dfrac{d}{dt}\displaystyle\int_\Omega N_3(x, t)dx = \int_\Omega g(u) N_2(x, t)dx. \end{cases} \qquad (7.18)$$

Let $\alpha_i(t) := \int_\Omega N_i(x, t)dx$. Obviously $\alpha_i(t) \geqslant 0$ for all $t > 0$. We define $\alpha(t) :=$
$\alpha_1(t) + \alpha_2(t) + \alpha_3(t)$. By the definition,

$$\frac{d}{dt}\alpha(t) = 0.$$

Thus, $\alpha(t) \equiv \alpha(0) = \int_\Omega [N_{1,0}(x) + N_{2,0}(x) + N_{3,0}(x)]dx, \forall t \geqslant 0$.

From the first equation of (7.18) we obtain $\alpha_1(t)$ is nonincreasing in t and $\alpha_1(t)$ is bounded below by 0. This yields the convergence

$$\alpha_1(t) \xrightarrow{t\to\infty} \alpha_1^\infty \geqslant 0.$$

From the last equation of (7.18) we obtain $\alpha_3(t)$ is nondecreasing in t and bounded above by $\alpha(0)$. Hence,

$$\alpha_3(t) \xrightarrow{t\to\infty} \alpha_3^\infty \geqslant 0.$$

Since $\alpha_2(t) = \alpha(0) - \alpha_1(t) - \alpha_3(t)$, we obtain $\alpha_2(t)$ convergence to $\alpha_2^\infty = \alpha(0) - \alpha_1^\infty - \alpha_3^\infty$. Analogously using Neumann boundary conditions one can obtain

$$\begin{cases} \int_0^\infty \left(\int_\Omega f(u)N_1(x,t) \right) dt \leqslant C_0 \\ \int_0^\infty \left(\int_\Omega g(u)N_2(x,t) \right) dt \leqslant C_0 \end{cases} \tag{7.19}$$

and

$$\int_0^\infty \|\nabla_x N_1(x,t)\|_{L^2(\Omega)}^2 dt \leqslant C_0, \quad \int_0^\infty \|\Delta_x N_1(x,t)\|_{L^2(\Omega)}^2 dt \leqslant C_0$$

where C_0 is a some positive constant. Based on estimates (7.19) we will study the first asymptotic behavior of subpopulations $N_i(x,t)$, $i = 1,2,3$. We start with $N_1(x,t)$. To this end, we decompose

$$N_1(x,t) = n_1(t) + N_1^\perp(x,t), \quad \text{where } N_1(x,t) \in H^\perp;$$

$$H^\perp := \left\{ w \in L^2(\Omega) \mid \int_\Omega w(x)dx = 0 \right\}.$$

Then for any $t > 0$

$$\int_\Omega N_1(x,t)dx = \int_\Omega n_1(t)dx = n_1(t)|\Omega|,$$

where $|\Omega|$ is denoted by the volume of bounded domain $\Omega \subset \mathbb{R}^n$.

Hence, $n_1(t) = \frac{1}{|\Omega|}\alpha_1(t)$. Therefore as $t \to \infty$

$$n_1(t) \xrightarrow{t\to\infty} n_1^\infty := \frac{1}{|\Omega|}\alpha_1^\infty. \tag{7.20}$$

Next we study asymptotics as $t \to \infty$ of $N_1^\perp(x, t)$. To this end, we multiply first equation of (7.17) by $-\Delta_x N_1$ and integrate over Ω. Then we get

$$\frac{1}{2}\frac{d}{dt}\|\nabla_x N_1(t)\|^2_{L^2(\Omega)} + d_3\|\Delta_x N_1(t)\|^2_{L^2(\Omega)} \leq \|f(u)N_1\|_{L^2(\Omega)} \cdot \|\Delta_x N_1\|_{L^2(\Omega)}$$

$$\leq \frac{d_3}{2}\|\Delta_x N_1(t)\|^2_{L^2(\Omega)} + \frac{1}{2d_3}\|f(u)N_1\|^2_{L^2(\Omega)}. \tag{7.21}$$

Therefore we get

$$\frac{1}{2}\frac{d}{dt}\|\nabla_x N_1(t)\|^2_{L^2(\Omega)} + \frac{d_3}{2}\|\Delta_x N_1(t)\|^2_{L^2(\Omega)} \leq \frac{1}{2d_3}\|f(u)N_1\|^2_{L^2(\Omega)}.$$

By Wirtinger inequality, we have

$$\|\nabla_x N_1(t)\|_{L^2(\Omega)} = \|\nabla_x N_1^\perp(t)\|_{L^2(\Omega)} \leq C_w\|\Delta_x N_1^\perp(t)\|_{L^2(\Omega)} = C_w\|\Delta_x N_1(t)\|_{L^2(\Omega)}. \tag{7.22}$$

Then from (7.21) and (7.22) we get

$$\frac{d}{dt}\|\nabla_x N_1(t)\|^2_{L^2(\Omega)} + \frac{d_3}{C_w}\|\nabla_x N_1(t)\|^2_{L^2(\Omega)} \leq \frac{1}{d_3}\|f(u)N_1(t)\|^2_{L^2(\Omega)}.$$

Since (7.19) implies $\|f(u)N_1(t)\|^2_{L^2(\Omega)} \in L^1(0, \infty)$ we conclude that

$$\|\nabla_x N_1^\perp(t)\|_{L^2(\Omega)} = \|\nabla_x N_1(t)\|_{L^2(\Omega)} \to 0 \text{ as } t \to \infty. \tag{7.23}$$

Thus, by (7.20) and (7.23) we get

$$N_1(x, t) \to n_1^\infty \text{ as } t \to \infty$$

strongly in $H^1(\Omega)$.

Next we study asymptotics of $N_2(x, t)$. To this end, we decompose

$$N_2(x, t) = n_2(t) + N_2^\perp(x, t),$$

where $N_2^\perp(x, t) \in H^\perp$. In the same manner as we did for $N_1(x, t)$ we obtain

$$n_2(t) = \frac{1}{|\Omega|}\alpha_2(t), \quad \alpha_2(t) := \int_\Omega N_2(x, t)dx$$

as well as

$$n_2(t) \to n_2^\infty := \frac{1}{|\Omega|}\alpha_2^\infty \text{ as } t \to \infty. \tag{7.24}$$

To study asymptotics $N_2(x, t)$ as $t \to \infty$ it remains to study $N_2^{\perp}(x, t)$ as $t \to \infty$. For these purposes, we multiply the second equation of (7.19) by $N_2^{\perp}(x, t)$ and integrate over Ω. This yields

$$\frac{1}{2}\frac{d}{dt}\|N_2^{\perp}(t)\|_{L^2(\Omega)}^2 + d_4\|\nabla_x N_2^{\perp}(t)\|_{L^2(\Omega)}^2 \leqslant \varepsilon\|N_2^{\perp}(t)\|_{L^2(\Omega)}^2$$

$$+ \frac{1}{4\varepsilon}\|f(u)N_2\|_{L^2(\Omega)}^2 - \int_{\Omega} g(u)N_2(x, t)(N_2(x, t) - n_2(t))dx. \qquad (7.25)$$

From (7.25) it follows that

$$\frac{1}{2}\frac{d}{dt}\|N_2^{\perp}(t)\|_{L^2(\Omega)}^2 + \frac{d_4}{2}\|\nabla_x N_2^{\perp}(t)\|_{L^2(\Omega)}^2 + \left(\frac{d_4}{2C_w^2} - \varepsilon\right)\|N_2^{\perp}(t)\|_{L^2(\Omega)}^2$$

$$+ \int_{\Omega} g(u)N_2^2(x, t)dx$$

$$\leqslant \int_{\Omega} g(u)u_2(t)N_2(x, t)dx + \frac{1}{4\varepsilon}\|f(u)N_2\|_{L^2(\Omega)}^2. \qquad (7.26)$$

By virtue of (7.24), we get

$$\sup_{t>0}\|u_2(t)\|_{L^\infty(\Omega)} \leqslant C_0.$$

Then by (7.26), we obtain

$$\int_0^\infty \int_{\Omega} g(u)u_2(t)N_2(x, t)dxdt \leqslant \int_0^\infty \|n_2(t)\|_{L^\infty(\Omega)} \cdot \|g(u)N_2(t)\|_{L^1}\, dt \leqslant \tilde{C}_0,$$

where $\tilde{C}_0$ is a some constant independent of t. By virtue of Proposition 4.2 based on the last inequality we obtain

$$\|N_2^{\perp}(t)\|_{L^2(\Omega)} \to 0 \text{ as } t \to \infty. \qquad (7.27)$$

Furthermore, integrating (7.26) over $[0, \infty)$ we have

$$\int_0^\infty \int_{\Omega} g(u)N_2^2(x, t)dxdt \leqslant \tilde{C}_0, \text{ hence } \int_0^\infty \|g(u)N_2(t)\|_{L^2(\Omega)}^2\, dt \leqslant \tilde{C}_0. \qquad (7.28)$$

Our next step is to obtain convergence of $N_2(x, t)$ as $t \to +\infty$ strongly in $H^1(\Omega)$. To this end, we multiply the second equation of (7.17) by $-\Delta_x N_2$. Then we get

$$\frac{1}{2}\frac{d}{dt}\|\nabla_x N_2(t)\|_{L^2(\Omega)}^2 + d_3\|\Delta_x N_2(t)\|_{L^2(\Omega)}^2 \leqslant \frac{d_3}{2}\|\Delta_x N_2(t)\|_{L^2(\Omega)}^2$$

$$+ \frac{1}{2d_3}\left(\|f(u)N_1(t)\|_{L^2(\Omega)}^2 + \|g(u)N_2(t)\|_{L^2(\Omega)}^2\right).$$

Hence by Wirtinger inequality (see Proposition 7.1) and taking into account $\nabla_x N_2(x, t) = \nabla_x N_2^\perp(x, t)$ we have

$$\frac{d}{dt}\|\nabla_x N_2(t)\|_{L^2(\Omega)}^2 + \frac{d_3}{2C_w^2}\|\nabla_x N_2(t)\|_{L^2(\Omega)}^2 + \frac{d_3}{2}\|\Delta_x N_2(t)\|_{L^2(\Omega)}^2$$

$$\leqslant \frac{2}{d_3}\left(\|f(u)N_1(t)\|_{L^2(\Omega)}^2 + \|g(u)N_2(t)\|_{L^2(\Omega)}^2\right).$$

Therefore in view of (7.19) and (7.28), we deduce from Corollary 4.1 that

$$\|\nabla_x N_2(t)\|_{L^2(\Omega)}^2 = \|\nabla_x N_2^\perp(t)\|_{L^2(\Omega)}^2 \to 0 \text{ as } t \to \infty \qquad (7.29)$$

as well as

$$\int_0^\infty \|\Delta_x N_2(t)\|_{L^2(\Omega)}^2 dt \leqslant \tilde{C}_0.$$

Thus, (7.24), (7.27), and (7.29) imply that

$$N_2(x, t) \to n_2^\infty \text{ as } t \to \infty \text{ strongly in } H^1(\Omega).$$

In the same manner we can study asymptotic behavior of $N_3(x, t)$ as $t \to +\infty$. Indeed, let

$$N_3(x, t) = n_3(t) + N_3^\perp(x, t)$$

where $N_3^\perp(x, t) \in H^\perp$. Then one can easily see that

$$n_3(t) \to n_3^\infty := \frac{1}{|\Omega|}\alpha_3^\infty \text{ as } t \to +\infty.$$

Multiplying the last equation of (7.17) by $-\Delta_x N_3$ we get

$$\frac{d}{dt}\|N_3(t)\|_{L^2(\Omega)}^2 + \frac{d_4}{2C_w^2}\|\nabla_x N_3(t)\|_{L^2(\Omega)}^2$$

$$+ \frac{d_4}{2}\|\Delta_x N_3(t)\|_{L^2(\Omega)}^2 \leqslant \frac{1}{d_4}\|g(u)N_2(t)\|_{L^2(\Omega)}^2.$$

Using exactly the same arguments as we did for $N_1(x, t)$ and $N_2(x, t)$ we obtain that

$$N_3(x, t) \to n_3^\infty \text{ strongly in } H^1(\Omega) \text{ as } t \to \infty$$

and

$$\int_0^\infty \|\Delta_x N_3(t)\|_{L^2(\Omega)}^2 dt \leqslant \tilde{C}_0.$$

It remains to obtain asymptotic behavior of calcium evolution $u(x,t)$, that is a solution of

$$\begin{cases} \partial_t u = d_1 \Delta_x u + d_2 g(u) N_2(x,t), \ x \in \Omega, \ t > 0 \\ \dfrac{\partial u}{\partial n}\bigg|_{\partial \Omega} = 0, \ u|_{t=0} = u_0(x). \end{cases} \tag{7.30}$$

Integrating (7.30) over Ω we get

$$\frac{\partial}{\partial t} \int_\Omega u(x,t)dx = d_2 \int_\Omega g(u) N_2(x,t)dx.$$

Hence

$$\int_\Omega u(x,t)dx = \int_\Omega u_0(x)dx + d_2 \int_0^t \left(\int_\Omega g(u) N_2(x,t)dx \right) dt.$$

On the other hand

$$\int_0^t \left(\int_\Omega g(u) N_2(x,t)dx \right) dt = \int_\Omega \left(\int_0^t g(u) N_2(x,t)dt \right) dx$$

$$= \alpha_3(t) - \alpha_3(0).$$

Let

$$u(x,t) = a(t) + \varphi^\perp(x,t),$$

where $\varphi^\perp \in H^\perp$. Then one can easily see (in the same manner as above) that

$$\alpha_3(t) \to \alpha_3^\infty \quad \text{and} \quad a(t) \to \underbrace{a^\infty + d_2(\alpha_3^\infty - \alpha_3(0))}_{\tilde{a}_\infty}, \ a^\infty := a(0) \text{ as } t \to \infty$$

and

$$\int_\Omega u(x,t)dx \to \int_\Omega u_0(x)dx + d_2(\alpha_3^\infty - \alpha_3(0)).$$

Multiplying equation for $u(x, t)$ in (7.30) by $-\Delta_x u$ we have

$$\frac{1}{2}\frac{d}{dt}\|\nabla_x u(t)\|^2_{L^2(\Omega)} + d_1\|\Delta_x u(t)\|^2_{L^2(\Omega)}$$

$$\leqslant \frac{d_1}{2}\|\Delta_x u(t)\|^2_{L^2(\Omega)} + \frac{d_2^2}{2d_1}\|g(u)N_2(t)\|^2_{L^2(\Omega)}.$$

Hence by Wirtinger inequality we get

$$\frac{d}{dt}\|\nabla_x u(t)\|^2_{L^2(\Omega)} + \frac{d_1}{4}\|\Delta_x u(t)\|^2_{L^2(\Omega)}$$

$$+ \frac{d_1}{4C_w^2}\|\Delta_x u(t)\|^2_{L^2(\Omega)} \leqslant \frac{d_2^2}{d_1}\|g(u)N_2(t)\|^2_{L^2(\Omega)}.$$

Taking into account that $\|g(u)N_2(t)\|^2_{L^2(\Omega)} \in L^1(0, \infty)$ and using Proposition 4.2 we obtain that

$$\|\nabla_x u(t)\|_{L^2(\Omega)} = \|\nabla_x \varphi^\perp(t)\|_{L^2(\Omega)} \to 0 \text{ as } t \to \infty. \tag{7.31}$$

Thus

$$u(x, t) \to \tilde{a}^\infty \text{ strongly in } H^1(\Omega) \text{ as } t \to \infty.$$

Furthermore, since we already obtained a priori bound for $\|\nabla_x N_2(t)\|_{L^2(\Omega)}$ (see (7.29)), then following word by word as it was in Sect. 4.2 we can derive boundedness of

$$\sup_{t \geqslant \delta}\|\Delta_x \varphi^\perp(t)\|_{L^2(\Omega)} = \sup_{t > \delta}\|\Delta_x u(t)\|_{L^2(\Omega)} \leqslant C_*.$$

From Lemma 4.1 we know for all $t \geqslant \delta, \theta \in (0, 1)$

$$\|u(t) - \tilde{a}^\infty\|_{L^\infty(\Omega)} \leqslant CC_\theta \sup_{t \geqslant \delta}\|\Delta_x \varphi^\perp(t)\|^{1-\theta}_{L^2(\Omega)} \cdot \|\nabla_x \varphi^\perp\|^\theta_{L^2(\Omega)}$$

and from the boundedness of $\Delta_x \varphi^\perp(t)$ it follows that

$$\|u(t) - \tilde{a}^\infty\|_{L^\infty(\Omega)} \leqslant \tilde{C}_*\|\nabla_x \varphi^\perp\|^\theta_{L^2(\Omega)}$$

for all $t \geqslant \delta$. With (7.31) we finally obtain

$$\|u(t) - \tilde{a}^\infty\|_{L^\infty(\Omega)} \to 0 \text{ as } t \to \infty \tag{7.32}$$

and the uniform convergence of $u(x, t)$ to the constant function $\tilde{a}^\infty$ is shown. Let us recall that $\tilde{a}^\infty := a(0) + d_2(\alpha_3^\infty - \alpha(0))$.

Remark 7.1 Due to the presence of Laplacian in the equations of subpopulations $N_i(x, t)$, we obtained (7.32) without structural assumptions as it was in Chap. 4 (PDE-ODE coupling).

Remark 7.2 As it was shown in Chap. 4, for Neumann BC, one can show that there exists $t_0 > 0$ and $\rho > 0$ such that

$$u(x, t) \geqslant \rho \text{ for all } t \geqslant t_0, \text{ a.e. } x \in \Omega.$$

Assume that $g(\rho) = g_\rho > 0$. Then multiplying the second equation of (7.17) by $N_2(x, t)$, we get

$$\frac{1}{2}\frac{d}{dt}\|N_2(t)\|^2_{L^2(\Omega)} + d_4\|\nabla_x N_2(t)\|^2_{L^2(\Omega)}$$
$$+ \int_\Omega g(u)(N_2(x, t))^2 dx \leqslant \int_\Omega f(u)N_1 N_2 dx. \qquad (7.33)$$

It follows from (7.33) that

$$\frac{1}{2}\frac{d}{dt}\|N_2(t)\|^2_{L^2(\Omega)} + g_\rho\|N_2(t)\|^2_{L^2(\Omega)} \leqslant \frac{g_\rho}{2}\|N_2(t)\|^2_{L^2(\Omega)} + \frac{1}{2g_\rho}\|f(u)N_1 N_2\|^2_{L^2(\Omega)}.$$

Since $\|f(u)N_1(t)\|_{L^2(\Omega)} \in L^1(0, \infty)$, then by Proposition 4.2 we obtain

$$\|N_2(t)\|_{L^2(\Omega)} \to 0 \text{ as } t \to \infty$$

and consequently $n_2^\infty = 0$.

Remark 7.3 Note that for Robin BC with $\alpha > 0$, we already know that $\exists t_0 > 0$ and $\exists \rho > 0$ such that $u(x, t) \geqslant \rho$ for all $t \geqslant t_0$ and a.e. $x \in \Omega$. Hence, as in Remark 7.2 one can show that $n_2^\infty = 0$ in the case of Robin BC.

Remark 7.4 As for the case for Dirichlet BC and Robin BC with $\alpha = 0$, under suitable condition on g, one can show that $n_2^\infty \neq 0$.

Remark 7.5 Analogous to PDE-ODE case one can prove partial and complete swelling depending on the boundary conditions as well as relation between α, c^-, and $\tilde{a}^\infty$.

7.3   Numerical Simulations

We illustrate the previous results on longtime behavior with numerical simulations in 1D, over the interval $x \in (0, 1)$. For this, we have to specify appropriate functions $f(u)$ and $g(u)$. Following [1–3] we choose

$$
f(u) = \begin{cases} 0, & 0 \leqslant u \leqslant C^-, \\ \frac{f^*}{2}\left(1 - \cos \frac{(u-C^-)\pi}{C^+-C^-}\right), & C^- \leqslant u \leqslant C^+, \\ f^*, & u > C^+, \end{cases}
$$

and

$$
g(u) = \begin{cases} \frac{g^*}{2}\left(1 - \cos \frac{u\pi}{C^+}\right), & 0 \leqslant u \leqslant C^+, \\ g^*, & u > C^+. \end{cases}
$$

The model parameters used are summarized in Table 7.1. They have been taken from our previous studies [1–3] and chosen primarily to support, demonstrate, and emphasize the mathematical results, not for quantitative prediction. We assume here that the diffusion coefficients are the same for all three classes of mitochondria, and that motility of mitochondria is smaller than diffusion of calcium ions.

The initial data for the calcium ion concentration are chosen such that at $x = 0$ the concentration is higher than C^+ and at $x = 1$ it is lower than C^-, connected by a cosine wave.

$$
u(x, 0) = \hat{C} \cdot (1 + \cos(x\pi)), \quad x \in (0, 1), \ \hat{C} = 250
$$

and

$$
N_1(x, 0) = 1, \quad N_2(x, 0) = 0, \quad N_3(x, 0) = 0, \quad x \in \Omega,
$$

i.e., we assume that initially swelling has not yet been initiated.

Table 7.1 Default parameter values, cf also [2] (with permission from Wiley)

Parameter	Symbol	Value	Remark
Lower (initiation) swelling threshold	C^-	20	
Upper (maximum) swelling threshold	C^+	200	
Maximum transition rate for $N_1 \to N_2$	f^*	1	
Maximum transition rate for $N_2 \to N_3$	g^*	1	
Feedback parameter	d	30	
Diffusion coefficient of calcium ions	d_1	0.2	
Diffusion coefficient of mitochondria	$d_m = d_{2,3,4}$	(Varied)	

All our simulations show nonnegativity of u, N_1, N_2, N_3 and that the solution converges to a spatially homogeneous steady state as $t \to \infty$. More specifically we find $N_1 \to 0$ $N_2 \to 0$, $N_3 \to 1$. With our assumption that the mitochondrial fractions have the same diffusion coefficients, we obtain from the model equations that $N := N_1 + N_2 + N_3$ satisfies the heat equation

$$\partial_t N = d_m \Delta_x N,$$

which, under our initial and boundary conditions, has the solution $N(x, t) \equiv 1$. Our numerical simulations satisfy this with at least 6 digits (data not shown).

In Fig. 7.1 we visualize the results of a typical simulation, where we choose for the diffusion coefficients of the mitochondria as $d_m = d_{2,3,4} = 0.02 < d_1$. The evolution of the calcium ion concentration is initially dominated by diffusion, leading to an obliteration of the spatial gradients that were introduced by the initial conditions. At about $t = 2.4$ it appears stratified, from where on the evolution is dominated by slight growth until steady state is reached. The calcium ion concentration gradients in the initial data lead immediately to gradients in the mitochondria distribution. The mitochondria fraction N_1 starts declining immediately, whereas N_2 and N_3 immediately increase. The rates that determine the swelling process depend on the calcium ion concentration which introduces gradients in the mitochondrial fractions. In the initial phase, where u is highest, N_1 is lowest and N_2 and N_3 are highest. The calcium ion concentration stratifies quickly, which induces also stratification of the mitochondrial populations, however, at a slower pace. Noteworthy is that between $= 12$ and $t = 2.4$ the profile of N_2 tips. Whereas in the earlier phase it was decreasing (as a function of x it becomes increasing later on, before it levels off and declines). This is due to the shift in balance between the transition of N_1 to N_2 and from N_2 to N_3.

In Fig. 7.2 we show the evolution of the space averaged quantities for the simulation that was visualized in Fig. 7.1. The fraction N_1 of mitochondria that have not yet engaged in the swelling process declines, and the fraction N_3 of completely swollen mitochondria increases. The fraction N_2 of mitochondria engaged in swelling first increases and then decreases. The amount of calcium ions increases and levels off eventually when the swelling process is complete. This resembles qualitatively the findings of the model without spatial movement of mitochondria in [1–3].

The difference between the PDE-PDE model that we study here and the PDE-ODE model of [1–3] is the diffusion of the mitochondria. To investigate the effect that this has we repeat the above simulation with $d_m = d_{3,4,5} = 0$, i.e., the case of the PDE-ODE coupled system of [1–3]. In Fig. 7.3 we show for selected time instances the solutions of the model with mitochondria diffusion vis-a-vis the corresponding solutions of the model without mitochondria diffusion. The differences between both solutions are only minor. In the mitochondrial fractions, the differences in N_1 and N_3 are largest. In these cases the mitochondria gradients are slightly higher in the case of the PDE-ODE model than in the case of the PDE-PDE model, as a consequence of Fickian diffusion obliterating gradients.

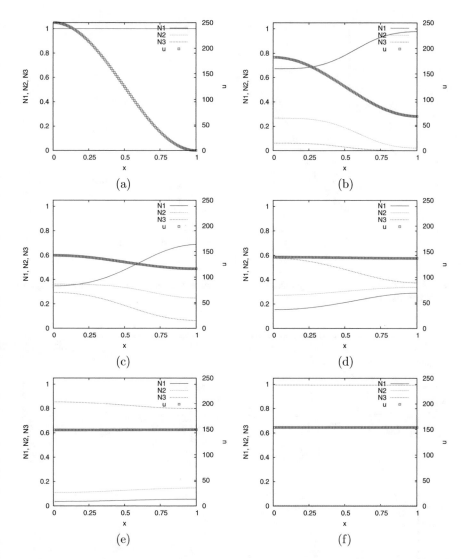

Fig. 7.1 Snapshot of model simulation at different time instances. Shown are the spatial profiles of the calcium ion concentration u (symbols), and of the mitochondrial fractions N_1, N_2, N_3 (solid lines). (a) $t = 0$. (b) $t = 0.4$. (c) $t = 1.2$. (d) $t = 2.4$. (e) $t = 4.4$. (f) $t = 9.0$ (with permission from Wiley)

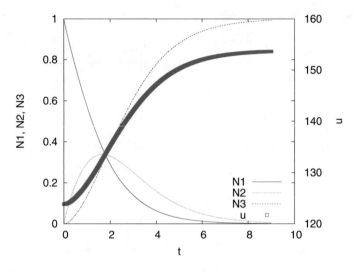

Fig. 7.2 Time evolution of the spatially averaged dependent variables for the simulation in figure (with permission from Wiley)

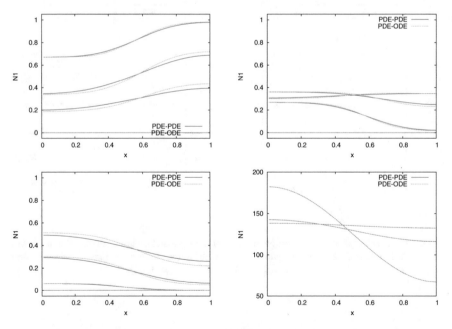

Fig. 7.3 Comparison of the solutions of the PDE-PDE model with diffusion of mitochondria with the results of the PDE-ODE model without mitochondrial movement. Plotted are the spatial profiles at times $t = 0.4, 1.2, 2.0$ (with permission from Wiley)

In the case of the calcium ion concentrations the differences are close to plotting accuracy. This suggests that (for the parameters tested here), the spatial gradients in the mitochondrial populations do not affect the spatial calcium ion distribution. In Fig. 7.3, the solutions of the PDE-ODE model are (slightly) larger in some places and (slightly) smaller in other places than the solutions of the PDE-PDE model. This suggest that the differences neutralize each other when the spatial averages of the dependent variables are taken. We verified this by comparing the average data as functions of time and found negligible differences between both models (data not shown).

Conclusion

Fickian diffusion obliterates gradients. Under Neumann boundary conditions the dynamics of mitochondrial swelling leads to homogeneous distributions of all components. Accounting for diffusive motility of mitochondria accelerates the spatial stratification but has only negligible effect on overall process dynamics, as expressed by spatial averages of mitochondrial fractions. For the parameters used in our simulations, spatial effects are controlled by the dynamics of calcium ions.

References

1. M. Efendiev, M. Ôtani, H.J. Eberl, A coupled PDE/ODE model of mitochondrial swelling: Large-time behavior of the homogeneous Dirichlet problem. J. Coupled Syst. Multiscale Dyn. **3**(2), 122–134 (2015)
2. S. Eisenhofer, A coupled system of ordinary and partial differential equations modeling the swelling of mitochondria. PhD thesis, Technische Universität München, 2013
3. S. Eisenhofer, M.A. Efendiev, M. Otani, S. Schulz, H. Zischka, On a ODE–PDE coupling model of the mitochondrial swelling process. Discrete Contin. Dyn. Syst. Ser. B **20**(4), 1031–1058 (2015)
4. K. Mitra, C. Wunder, B. Roysam, G. Lin, J. Lippincott-Schwartz, A hyperfused mitochondrial state achieved at G1–S regulates cyclin E buildup and entry into S phase. Proc. Natl. Acad. Sci. **106**(29), 11960–11965 (2009)

Printed in the United States
By Bookmasters